AN INTRODUCTION TO POLITICAL GEOGRAPHY

Old powers are falling. New states are emerging. The gap between East and West is narrowing. Yet the developments in the Middle East and Eastern bloc, the increasing disparity between the rich and poor nations, the intensification of economic competition between former political allies in the rich core, pose new threats and tensions for the New World.

An Introduction to Political Geography, in its first edition, helped to shape the study of the discipline. Entirely revised and updated this new edition explores political and geographic change within the same accessible framework, emphasizing the need for a fluid approach to the study of the international order, the nation-state, as well as social movements.

Examining the North–South and East–West dimensions in the World Order and the rise of new centres of power from an historical perspective, Part I provides a background for discussion of current trends and future developments. The nation state, the key unit that binds the generality of world order with the particularity of individual households, is introduced through analytic study in Part II, whilst Part III utilises detailed case studies to discuss social movements and the politics of time and place.

Entirely revised and updated this new edition emphasizes the trend towards globalization but challenges the traditional integration of the world systems approach. A new section on the political geography of participation considers the concept of the global village, with its concerns for global justice and environmentalism. The extent to which active participation of people can determine social and political change prompts a range of original discussions.

John Rennie Short is Professor of Geography at Syracuse University, New York.

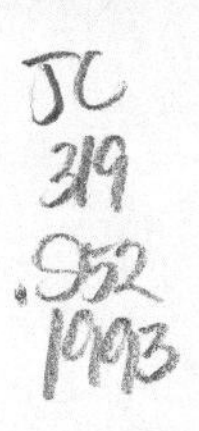

AN INTRODUCTION TO POLITICAL GEOGRAPHY

Second edition

John Rennie Short

London and New York

First published 1993
by Routledge
11 New Fetter Lane, London EC4P 4EE

Simultaneously published in the USA and Canada
by Routledge
a division of Routledge, Chapman and Hall, Inc.
29 West 35th Street, New York, NY 10001

Typeset in 10pt September by Leaper & Gard Ltd, Bristol
Printed in England by Clays Ltd, St Ives plc

A catalogue reference for this title is available from the British Library
ISBN 0-415-08226-9
0-415-08227-7

Library of Congress Cataloging in Publication Data
Short, John R.
An introduction to political geography / John Rennie Short. –
2. ed.
p. cm.
Includes bibliographical references and index.
ISBN 0-415-08226-9. – ISBN 0-415-08227-7
1. Political geography. I. Title.
JC319.S52 1993
320.1'2–dc20 92-24742
CIP

CONTENTS

FOR MY BROTHER, KEVIN

FIGURES

TABLES

ERRATA

Page 24 Figure 1.9	Rent commodity prices *should read* Real commodity prices
Page 64 Figure 3.4	*coded tints for the Netherlands and Denmark should be transposed*
Page 127 Table N.1	*second column should read* 1968 (*not 21968*)
Page 152 Table 9.1	after Castells, 1983

PREFACE TO THE SECOND EDITION

A second edition of this book is long overdue. I wrote the first in the autumn and winter of 1979 and the spring and summer of 1980. The book was a necessity. I had been given a lectureship at Reading University in 1978 and was expected to teach political geography; this was the teaching course of the person who had retired. He left to go on a world cruise and I was lumbered with teaching a course for which I had no experience. I looked round the library. I was disappointed. There seemed nothing exciting or new in the political geography section, just tired texts on boundaries and frontiers. With the arrogance of youth and the confidence of the ignorant I decided to write my own text.

The result was the first edition. My students seemed to like the course and the reviews were, on the whole, favourable. On the whole is one way of saying that there were some real stinkers. A Canadian accused me of being anti-American, some thought I was too Marxist, others not Marxist enough, some said too theoretical, others not theoretical enough. Reading the reviews made me realize that you cannot please all the people all the time. All you can aim for is a measure of intellectual honesty and as high a degree of integrity as possible.

A second edition gives me a number of opportunities. It allows me to bring the narrative sections up to date. The world has moved on and major changes have occurred especially in East–West relations. It also allows me to correct a number of errors. My favourite was on page 12, where a misprint had the Spanish first colonizing the islands of the Caribbean in 1934. One humourless individual wrote to me, again from Canada, criticizing me for my lack of historical knowledge. A new edition gives me the chance to update the first third of the book and improve the last two thirds. Part I keeps its basic shape, while Parts II and III have been recast. I have tried to keep the good points of the first edition while correcting mistakes, rewriting entire chapters and improving the whole exposition.

Note

This book was written at a time of rapid change in the political order. I have tried to include all the major changes. However, by the time you read this, there will have been further developments, which may have affected national boundaries. The book reflects the state of the world in September 1992.

ACKNOWLEDGEMENTS

I have to thank a number of people who helped the second edition see the light of day. Heather Browning drew the maps and Mandy Jeffery typed a number of drafts of the entire manuscript, I owe her a very special thanks. David Knight made extensive comments in an earlier version. His comments were argumentative, stimulating and, ultimately, very generous.

INTRODUCTION

> The attempt is to find concepts and methods which cast light on the real world instead of codifying it into the obsolete categories of the academic establishment
>
> (Castells, 1980)

AIMS AND OBJECTIVES

A full understanding of society can only be achieved through analysing the strands which link spatial structures, political processes and economic systems. The overall aim of the book is to analyse these connecting strands.

The traditional subject matter of geography – the relationships between people and nature, people and space, people and places – cannot be separated from political considerations. The explicit focus on these considerations constitutes the general subject matter of political geography.

THE BACKGROUND

Peter Haggett once described geography as a Los Angeles of an academic city: all sprawling neighbourhoods and no centre. Like most big cities the neighbourhoods change over time, some retaining their exclusivity while others, once fashionable, become rundown and dilapidated. Political geography is an inner-city neighbourhood. It was important in the early development of a discipline but was bypassed by the growth of new suburbs. More recently, however, it has become fashionable again. Political geography is in the process of academic gentrification.

In the nineteenth and early twentieth centuries geography had a very strong political focus. Many of the early geographers such as Peter Kropotkin, Sir Halford Mackinder and Isaiah Bowman were explicitly concerned with the relations between politics and geography in both their published work and their public lives. Mackinder, for example, was a British MP, a High Commissioner in Russia and chairman of various government committees. Bowman was an adviser to US President Wilson at the Versailles Peace Treaty meetings. Sadly, this concern with the very stuff of politics waned after Mackinder and Bowman. Geo-politics became discredited and political geography became an ossified subdiscipline of a tired subject, often taught, never researched, a prisoner to outdated theories.

The surge of the new geography of the 1950s and 1960s bypassed political geography. With spatial analysis as its theme, neo-classical economics as its accounting frame, and logical positivism as its methodological underpinning, it could not accommodate a political geography. The emphasis of neo-classical economics on the economy as a harmonious, self-regulating system, where each factor of production receives its fair reward, ignored questions of conflict and inequitable distribution. The focus of logical positivism directed attention to verifiable empirical statements in particular and data analysis in general and away from the operation of the more incorporeal power relations within society. A truly political geography could not flourish in such a climate. The explicit analysis of politics was being taken over by another social science discipline, political science; a discipline which, in the words of Cobban (1953), was a device for avoiding politics without achieving science. Largely ignored by its discipline and lacking much theoretical sustenance from political science, it is little wonder that political geography was a moribund subject.

Things began to change in the 1960s and 1970s. The end of the post-war economic boom was reflected by the social sciences in the growth of approaches which focused on power, conflict and the inequitable distribution of life chances and resources. The change occurred in many disciplines – witness the resurgence of Marxist economies and the growth of radical sociology – and was seen in the emergence of interdisciplinary approaches which did not accept the artificial demarcation of knowledge suggested by traditional academic disciplines. Human geography was affected by these changes. The emergence of a radical geography and the development of a critical awareness amongst geographers led to important questions of 'Who gets what?' and 'Why does who get what?'

The result was a new political geography, which consists of old topics re-examined and new areas of enquiry. This rekindled interest is evident in the growing number of textbooks devoted to the subject, and the establishment of such journals as *Political Geography Quarterly* in 1982 and *Environment and Planning C: Government and Policy* in 1983. In this book we will concentrate on three broad areas of interest of this new political geography: the international order, the nation-state and social movements.

STRUCTURE OF THE BOOK

Part I examines the world order. Chapter 1 looks at the North–South dimension of the world order and Chapter 2 considers the East–West split. A historical approach is taken in order to provide a sound background to an examination of current trends and possible future developments. We need to look into the past to see the present.

In Part II the notion of the nation–state is introduced. The state is a key unit of analysis; it provides the explanatory glue that binds an understanding of the world order with an analysis of events at the local level.

Part III discusses the political geography of social movements and provides a discussion of the politics of location and the politics of place.

The three parts of the book have a different approach as well as different subject matter. Part I is an historical exposition, Part II is more analytical while Part III contains detailed case studies. A comprehensive political geography should have room for all three approaches.

GUIDE TO FURTHER READING

Books quickly date. To keep up with recent developments in political geography keep an eye on two journals – *Political Geography* and *Environment and Planning C: Government and Policy*. There is also an annual survey of the subject in *Progress in Human Geography*. The writers change every three years so there is a variety of opinions. Also have a look at *World Government* (Oxford University Press, New York) edited by Peter Taylor.

Works cited in this chapter

Castells, M. (1980) 'Cities and regions beyond the crisis: invitation to a debate', *International Journal of Urban and Regional Research* 4, 127–9.

Cobban, A. (1953) 'The decline of political theory', *Political Science Quarterly* LXVIII, 321–37.

Part I

THE POLITICAL GEOGRAPHY OF THE WORLD ORDER

This section is concerned with the political geography of the world order. This is a huge, complicated topic. We can, however, simplify the picture by identifying a coarse-grained grid. Two sets of co-ordinates are: the North–South division, as it is now known, of rich and poor countries and the East–West split, which used to divide the capitalist and socialist blocs. Since the late 1980s the East–West division has become less relevant, but it is still important to see its evolution if we are to understand its disappearance.

Chapter 1 considers the North–South division, chapter 2 examines the East–West split and chapter 3 looks at the rise of new centres of power.

1

UNEVEN DEVELOPMENT: THE CAPITALIST WHIRLPOOL

In the newly opened up countries the capital imported into them intensifies antagonisms and excites against the intruders the constantly growing resistance of the peoples who are awakening to national consciousness; this resistance can easily develop into dangerous measures against foreign capital. The old social relations become completely revolutionized, the age-long agrarian isolation of 'nations without history' is destroyed and they are drawn into the capitalist whirlpool.

(Hilferding, 1910)

INTRODUCTION

Figure 1.1 shows a range of data for Bangladesh and the USA. Behind these figures lie vastly

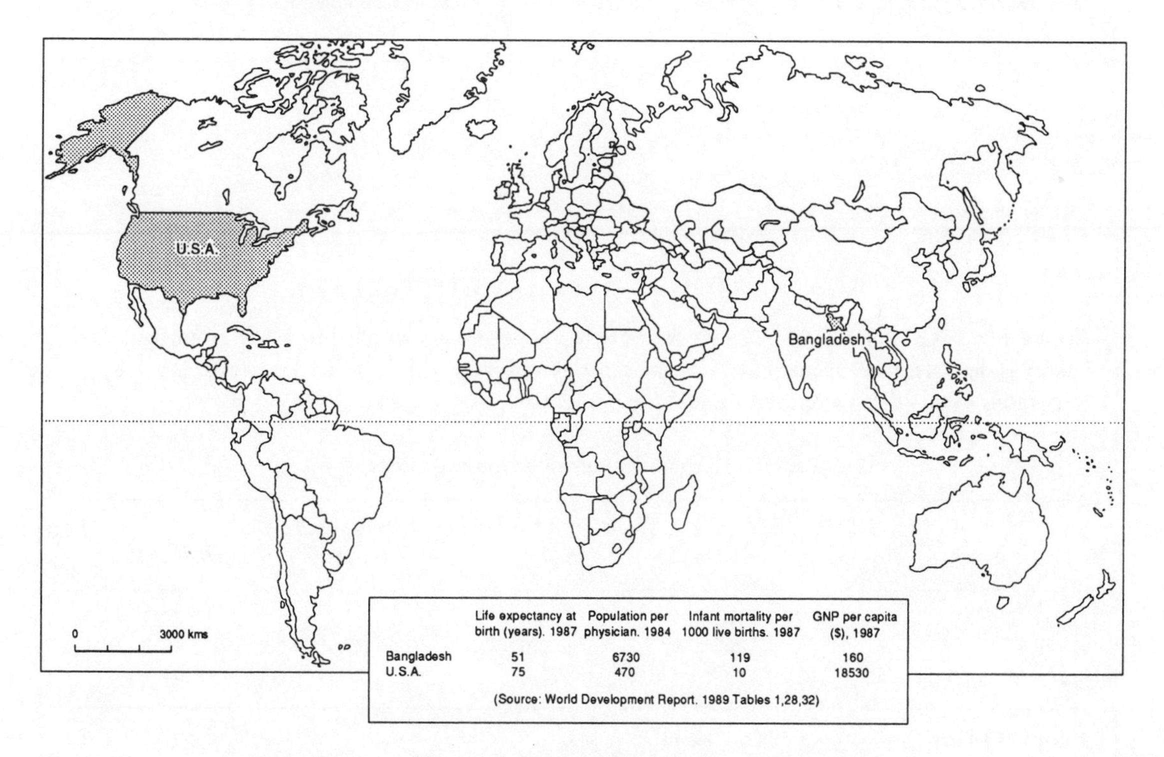

	Life expectancy at birth (years). 1987	Population per physician. 1984	Infant mortality per 1000 live births. 1987	GNP per capita ($), 1987
Bangladesh	51	6730	119	160
U.S.A.	75	470	10	18530

(Source: World Development Report. 1989 Tables 1,28,32)

Figure 1.1 Comparative statistics

BOX A: THE ROSTOW MODEL

In outline the Rostow model identifies five stages to economic growth:

1 *Traditional society*: characterized by limited technology, static social structure.
2 *Precondition take-off*: rise in rate of productive investment and evolution of new elites
3 *Take-off*: marked rise in rate of productive investment, development of substantial manufacturing sectors and emergence of new social and political frameworks which encourage and aid sustained economic growth.
4 *Drive to maturity*: impact of growth affects the whole economy.
5 *Age of high mass-consumption*: shift toward consumer durables.

The model has been used to describe the experience of selected countries (see Figure A.1). Criticisms of the model have centred on its inability to identify casual mechanisms between the different stages.

References

Baran, P. A. and Hobsbawm, E. J. (1961) 'The stages of economic growth', *Kyklos* 14, 324–42.
Rostow, W. W. (1960) *Stages of Economic Growth: A Non-Communist Manifesto.* Cambridge University Press, Cambridge.
Rostow, W. W. (1978) *The World Economy.* University of Texas Press, Austin and Landin.

BOX B: RICH AND POOR COUNTRIES

We have to be careful when we use the terms rich and poor when applied to whole countries. The reality is that there are rich people in poor countries and poor people in rich countries. Income is very rarely evenly distributed throughout the population.

Table B.1 Income distribution in three countries

		Percentage share of income by percentile group		
	Year	*Highest 5%*	*Highest 20%*	*Highest 10%*
Bangladesh	1981–2	6.6	45.3	29.5
Brazil	1977	2.0	66.6	50.6
USA	1980	5.3	39.9	23.3

Source: World Development Report 1989, Table 30.

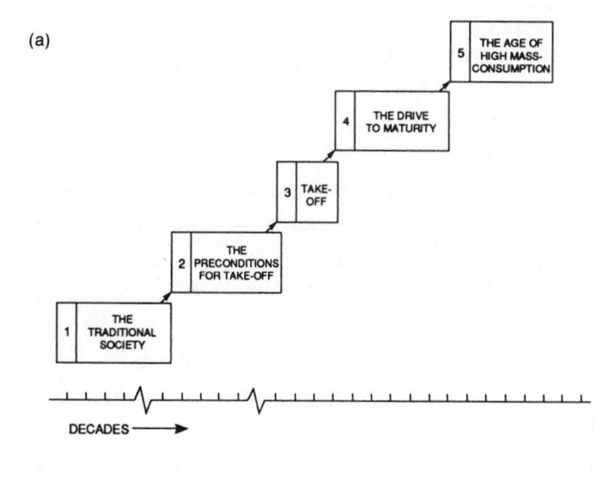

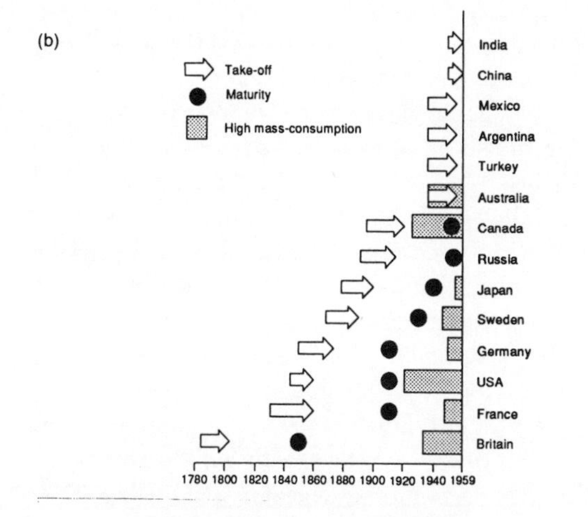

Figure A.1 The Rostow model

different life experiences. In Bangladesh most people rarely live beyond 50, they have to share a physician with over 6000 other people and over 1 in 10 of their children die before reaching the age of 1. Compare that with the USA where the average person lives to be 75, most people have better access to medical care and are less likely to experience the agony of the death of a child.

These figures are the tangible effects of an *uneven development* in the world. *Uneven* in the sense of marked inequalities, *development* in the broad sense of economic growth and social progress. Bangladesh and the USA are at the two extremes of the continuum of uneven development, one very poor, the other very rich. They highlight the basic division of the world into rich and poor countries. This division has been given a variety of names – developed–undeveloped, First World–Third World, North–South. Whatever the labels used the basic question remains. How did it come about?

One model used to explain this variation has been put forward by the US economist W. W. Rostow. He argues that economic growth has been achieved by only a few countries. In the rest, however, once the barriers of traditional society are broken down, growth can be achieved, leading to the final stage of high mass-consumption: left to market forces and time all countries will look like the USA. Underpinning the model is the notion that the international trading system is a harmonious system in which all countries can benefit. The crucial policy questions of the *Rostow model* are: what are the best methods of establishing the preconditions for growth and how can economic take-off be achieved? Poverty in the Rostow view of things results from lack of involvement in world trade.

The alternative model, the *core–periphery model*, takes the very opposite stance: poverty arises from involvement with the world economy. The core–periphery model refers to the spatial division of the world and a set of economic relationships. The core consists of the rich nations – Japan, North America and Europe while the periphery consists of all other countries. This broad-scale division of the world economy marks off rich countries from poor, economically advanced from economically underdeveloped, dependent from relatively independent economies and economic development as growth from economic development as growing dependence. The rich countries of the 'north' constitute the core because the international economy revolves around them and they have been the moving force guiding the development of an integrated world economy. The 'south' is peripheral in the sense that its economies are articulated to the needs of the rich core countries and the pace and character of its development have been shaped by contact with the core. There are, of course, elaborations to this simple structure:

- We can identify developed peripheral economies such as Australia and South Africa where the extraction of minerals and the production of primary commodities has been associated with relatively high rates of internal economic growth and high average incomes.
- We can also identify semi-peripheral countries which have achieved some level of autonomous economic growth. Two types can be identified: countries such as Brazil, which have been moving in the direction of periphery to core, and countries such as Britain, which seem to be moving in the opposite direction, from core to periphery.

THE HISTORICAL BACKGROUND

Beginnings of the world economy

The growth of the core–periphery structure of the world economy can be said to have started around 1500. That date marks the beginning of European overseas expansion and the articulation of an international economy based on flows of goods and people to and from Europe. The foundations for this expansion had been laid almost 350 years earlier. From about 1150 the increase in trade both within Europe and between Europe and the East saw the development of a merchant class, the growth of urban trading centres – Genoa, Venice, Naples and Milan in the south, the Hanseatic League towns of Hamburg, Lubeck and Bremen in the north – and the beginnings of money being used as a means of exchange. The merchants, the urban trading circuits and the use of money were essential prerequisites for the development of long-distance trade.

Between 1500 and 1600 European overseas expansion emanated from Iberia. The search for bullion, the demand for spices and the need for fuel (mainly wood) and food all gave impetus to exploration and when the Ottoman Empire obstructed overland trade with Africa and Asia, Iberia was well placed as a base for forging new trading links with the East. Portuguese merchants strung a number of trading posts along the main sea routes to Africa, India and the Far East and ships from Lisbon sailed regularly to Goa (India), Colombo (Sri Lanka), Malacca and Macao, collecting pepper, cinnamon, cloves and nutmeg. Contact in the East was limited to trade with the ports. The most important developments in colonization took place in the Americas. The Spanish first colonized the islands of the Caribbean in the late fifteenth century. Following the discoveries made by Cortes and Pizarro they then moved into Mexico and Peru. The Portuguese meanwhile obtained the eastern half of Brazil after the Papal settlement of 1493, which divided the New World between two Old World powers.

Throughout the sixteenth century the Americas were the scene of extensive exploitation as sugar plantations were created and the land was plundered for gold and silver. Both the mines and plantations used slave labour. Initially, the indigenous Indians had been used but the appalling fatalities and the increasing demand led to the growth of the slave trade. A rudimentary triangular trade was developed, with ships sailing from Europe to West Africa, taking slaves to the Americas and then returning to Europe with bullion and sugar. The sixteenth century thus witnessed the growth of the core–periphery structure from its early embryonic form. With the slave trade and the colonization of the Americas more of the world was brought within the orbit of Europe's trade. There were also the beginnings of the international division of labour, with slave labour in the Americas providing primary commodities for European markets. The gold and silver from Central America which flowed through the trading arteries of European merchant cities allowed the accumulation of money and commercial capital which eventually was to lay the basis for further economic growth. The plunder from the Americas provided the basis for investment in the early manufacturing industries of Europe. The nature and the timing of the incorporation of American territories into the European world economy was to determine the subsequent development of these areas.

By the beginning of the seventeenth century both Spain and Portugal were being hard pressed by other European powers. From 1600 onwards the power of Spain and Portugal began to decline as the struggle for commercial mastery was disputed between Holland, Britain and France. The centre of gravity in the core was beginning to shift.

The sharpening of conflict arose as the suckers of trade, which stretched out from the various countries, overlapped and struggled for sustenance in the same areas of the world. The conflict was not restricted to the commercial rivalry between the great trading companies. Commercial rivalry escalated into conflicts between states as each nation began, in varying degrees, to accept the *mercantilist idea* that it was the business of the state to promote the economic interests of the country and that this could best be achieved by stimulating foreign trade. For the mercantilists, foreign trade was seen as the chief method of increasing national wealth. Initially, it was thought that the best way to increase wealth was to accumulate gold and silver; the Spanish overseas expansion in the sixteenth century was partly grounded in this belief. Eventually, the emphasis changed to a belief in the efficacy of the import of raw materials and the export of manufactured goods. The state was called upon to achieve conditions favourable to the balance of trade, for as Thomas Munn counselled in 1622: 'We must ever observe this rule: to sell more to strangers yearly than we consume of theirs in value' (quoted in Rude, 1972, p. 267). Favourable conditions of trade meant that the continued supply of raw materials had to be assured and markets for finished goods had to be made safe from foreign competition.

The mercantilists believed that the world's total wealth was fixed in quantity like a giant wealth cake and could not be increased; any increase in one nation's slice of wealth could only be achieved at the expense of others. It thus followed that the most favourable conditions of trade were achieved in colonial empires where the terms of trade were shaped for the benefit of the merchants of the (metropolitan) country. For the mercantilists, colonies were simply branch plants providing valuable commodities and offering secure markets. The overseas expansion of the core states in the seventeenth and eighteenth centuries was to all intents and purposes a commercial undertaking – an undertaking whose aim was to achieve a self-sufficient economic empire and whose driving force was mercantile capitalist.

The pace of colonial expansion and the size of colonial holdings were shaped by the changing balance of forces between the core countries and their commitment to overseas expansion. In the first half of the seventeenth century the Dutch were the most dynamic nation, their ships spinning a web of trading links as far as Indonesia, India and North and South America. The efficiency and aggression of the Dutch trading machine were the envy of Europe and when Louis XIV's minister Colbert attempted to invigorate the French trading companies he modelled them on the Dutch. The solid faces which peer out from the canvases of Rembrandt, Vermeer and Hals belonged to the most aggressive merchants of their time. The success of the Dutch attracted the retaliatory action of other nations. In 1651 republican England passed the Navigation Act, which laid down the principle that trade with the colonies should only be carried on by English ships. The act was specially designed to defeat the Dutch hold over the shipping trade and to direct the valuable entrepot trade (the re-export of colonial commodities to Europe) towards England. The wars with Holland which followed this act and which lasted on and off from 1652 to 1674 were the military expression of the commercial rivalry between the two states. The ultimate English victory allowed English merchants to establish trading links with the Far East and India.

By the end of the seventeenth century Spain, Holland and Portugal were no longer major actors in the struggle for power; the stage was dominated by Britain (England joined with Scotland in the 1707 Act of Union) and France. Merchants in both countries were expanding their overseas operations and the two commercial empires faced each other in North America, the West Indies and India. The pivotal position was held by the West Indies. Throughout the eighteenth century, trade with the West Indies was the most valuable of all colonial trade. France held the lucrative sugar plantation islands of Guadeloupe and Martinique and the triangular trade of slaves from Africa, and the return cargo of sugar from the Indies to Europe constituted about a quarter of all France's commercial operations. This triangle of trade was even more important for Britain. Since 1713 Britain had had monopoly control over the supply of slaves to the Spanish American empire and almost a third of Britain's commercial operations consisted of the trade in manufactured goods and rum to Africa, slaves from Africa to the Americas and the transfer of sugar, cotton and tobacco from the Americas to Britain. The fine buildings and wealthy merchants of Liverpool, Glasgow and Bristol were solid testimony to the value of the Atlantic trade.

The conflict between France and Britain which had its roots in commercial rivalry took on greater substance in North America, where the dispute over claims to the huge hinterland became a major cause of the Seven Years War (1756–63). Britain's victory marked the beginnings of British pre-eminence in the colonial stakes. As Williams notes:

> Of the imperial powers of Europe which had fought for dominance in the eighteenth century Britain emerged supreme. The loss of the American colonies was more than counterbalanced by gains in other parts of the world. Britain's merchants dominated the trade routes in time of peace; her armed fleets controlled them in time of war. With her massive financial and industrial resources Britain was able to engage in full-scale war against France on the continent, and at the same time enlarge her colonial and commercial empire overseas. By 1815 Britain's imperial predominance was unchallenged, and the nation had entered an era of unprecedented growth in which her overseas empire and trade played an indispensable part.
>
> (Williams, 1966, p. 188)

Industrial capitalism, empire and neo-imperialism

The nineteenth century marked a change in the core–periphery structure. One country was to dominate the core while the core–periphery relationship was to develop from one of plunder and transfer of primary commodities into the purchase in certain peripheral areas of goods produced in the core. The story of the core–periphery in the first two-thirds of the nineteenth century is essentially the history of British hegemony in industrial production, colonial expansion and international trading. Britain was the first country to experience the 'qualitative and fundamental transformation' of the Industrial Revolution (Hobsbawm, 1968). Two distinct phases of industrialization can be noted. In the first period, from 1780 to 1840, cotton manufacturing was the leading sector whereas coal and iron took that role in the second period, from 1840 to 1900. In both periods a healthy export trade with the periphery, in association with a state prepared to formulate foreign policy to economic ends, were influential.

The cotton trade was intimately associated with Britain's empire. The raw material initially came from the West Indies and North America while the cotton cloth and garments were sold to the colonial markets. At the zenith of cotton's importance in 1830 one-half of all British exports consisted of cotton products and over three-quarters of these cotton goods went to the colonies in Africa, the West Indies and India. By 1850 India alone was absorbing 25 per cent of all Lancashire's cotton exports. This large and expanding export market provided the stimulus to continued production of cotton even when the domestic market suffered slumps and depressions. Britain's large export market gave an extra boost to domestic production. Britain's empire arose from the needs of commerce and industry while the empire helped create Britain's industry and commerce.

In the second phase of industrialization the pace was set by the exploitation of coal and the manufacture of iron and steel. Coal provided the lifeblood, and iron and steel the railway arteries for this later phase of economic growth. Like the first, the second phase relied upon the continued demand from foreign, especially colonial, markets. By 1850 over 80 per cent of the iron and steel produced in Britain was exported abroad, mainly for the construction of railways. The overseas railway boom provided new investment opportunities for British capital. In 1870 over 10 per cent of all British overseas investment was in Indian railways, providing a return of 5 per cent.

A distinction can be drawn in British trading arrangements between the *formal*, *informal* and *semi-formal* empires.

- The formal empire was that part of the British trading system which was also under direct political control.
- The informal empire was that constellation of countries and areas which were tied to Britain by trade and other economic links but with little or no explicit political connections. The distinction between the formal and informal empires is the difference between imperialism and neo-imperialism.
- The semi-formal empire consisted of countries such as Persia and Egypt under the Khedive Ismail, whose finances and/or trading arrangements were in the hands of Britain, but which were not officially declared colonies.

The formal empire arose from the need to integrate countries into the British trading system and to safeguard those areas bordering the major trading routes. India was the jewel in the crown of the British formal empire. Even the most liberal of nineteenth-century thinkers and the most committed free-trader could scarcely consider Britain without India. The British presence in India was initially restricted to the areas around Bombay, Madras and Bengal. Throughout the early nineteenth century Britain annexed the native states bordering the initial footholds and signed treaties with the more distant states. These treaties were often the first indication of future annexation. The Indian film director Satyajit Ray has brilliantly chronicled the annexation of Oudh in his film *The Chess Players* (1977). The subsequent story is one of the Indian economy being made subservient to British interests. Taxes, and especially the land tax, were to pay for the British presence in India, including the cost of the civil administration. The cost of military expenditure outside India, such as the Afghan war of 1839–42, was also charged to the Indian revenue account. This lootish direction of revenue is estimated to have amounted to £27 million (in 1865 prices) by 1870.

The terms of trade were manipulated to foster the sale of British manufactured goods. In the earlier decades of the nineteenth century India had a flourishing cottage industry of cotton manufacturing centring on Dacca. Trade relations were structured in a deliberate policy of de-industrialization. The duty paid in India for British silk and cotton goods was only 2 per cent while the British tariff on silk and cotton goods made in India was 10 per cent. Encouragement was also given to the Indian production of commodities necessary for British industry. Less tax was levied on land given over to cotton production than any other crop. The result was twofold. India began to supply more of Britain's raw cotton imports and, since less land was given over to food production, the poorer people suffered more from periodic famine and food shortages. The economic history of British involvement in India contains all the elements of the general core–periphery relationship that emerged around the world in the nineteenth and early twentieth centuries.

India became an integral part of the British economy as a source of raw materials, a purchaser of manufactured goods and a site for large-scale capital investment. Subsequent Indian development was shaped by this colonial experience. Other parts of the formal empire often turned out to be less attractive propositions. The incorporation of other, less wealthy areas into the British empire can be seen as miscalculations, speculative ventures proved wrong or attempts to control the

BOX C: HISTORICAL SUMMARY OF CORE-PERIPHERY RELATIONS

A thumbnail sketch of core–periphery relations up until 1945 would highlight three major points:

1. From the sixteenth to the nineteenth century the general story is one of incorporation of much of the world into the core's sphere of influence. Much of the world's history is a product of this incorporation. The incorporation involved a variety of methods from colonization, the indirect control of informal imperialism through the control of collaborative elites, to the direct control of formal imperialism which increased in the last half of the nineteenth century.
2. This incorporation involved a definite division of labour. Initially the periphery provided a source for raw materials, then it developed as a site for capital investment and as a market for goods produced in the core. This relationship laid the basis for underdevelopment because there was unequal exchange of cheap raw materials and expensive manufactured goods. The terms of trade laid the basis for the accumulation of wealth and economic growth in the core. The industrial revolution in Britain, for example, owed much of its origins to the colonies which were a source of cheap raw materials and a large secure market for manufactured goods.
3. Most of the conflicts in Europe were caused by the battle for supremacy of the core.

Century	*Competing powers*
16th	Spain, Portugal
17th	Spain, Netherlands, England, France
18th	Britain, France
19th	Britain

In the periphery, resistance to core domination fuelled a whole variety of liberation struggles.

trading routes to and from the richer parts of the empire.

The history of British colonial expansion in Africa, in contrast, has been termed a giant footnote to the history of British control in India. Although this may overstate the case, an important element of British foreign policy in Africa can be seen as a concern for the security of trading routes to India.

Imperialism could be a costly business. There were many Victorians who considered overseas possessions too great an expense. Gladstone expected the formal empire to crumble and Disraeli hoped that it would. The Utilitarians, Bentham and Mill, considered colonies a burden and the Manchester School continued to preach the benefits of free trade. In one sense the critics were correct. Imperialism and the existence of the formal empire were a sign of failure, namely the inability to integrate an area or country into the British trading system without imposing political, military and administrative control.

The economic success story of British capitalism in the nineteenth century was not the formal empire but the informal one. In Latin America, for example, imperialism scarcely seemed worthwhile when investments in trade, railways, shipping and public utility companies were receiving rates of return almost 2 per cent higher than the yield in British colonial government securities (Platt, 1977). Latin America was incorporated into the world economy but not into the formal empires of

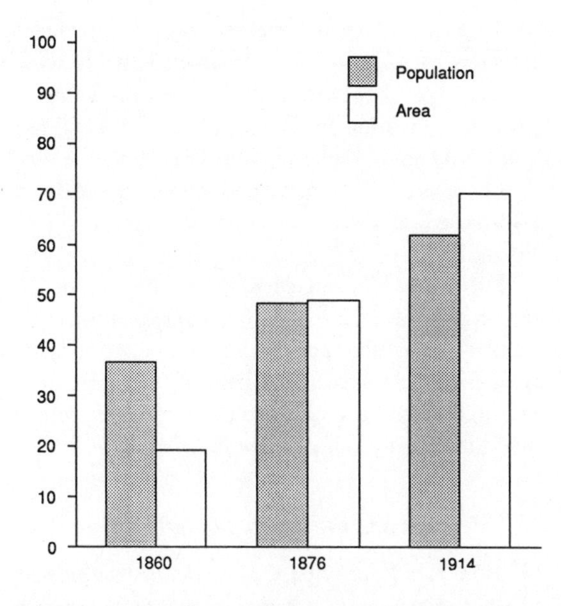

Figure 1.2 The expansion of European formal empires: Europe and colonies as a percentage of world surface area and population

core states because there was no need for formal imperialism. The ruling elites of Latin America aided and desired the involvement of foreign capital. British interests in Latin America were huge and ranged from shipping and railway companies, banks and insurance companies, to control of the Peruvian *guano* and Chilean nitrate industries and ownership of the many utility companies supplying urban transport and piped water. Incorporation into the world economy produced the typical pattern of Latin America providing the raw materials while Britain, and later the other core states, sold manufactured goods and invested the capital necessary for the exploitation and transport of these raw materials and primary commodities. By 1913, £999 million, roughly a quarter of all British capital investment overseas, was invested in Latin America with almost a quarter of this total being invested in Argentine railways. The form of this economic development was to mould the economies of Latin America to the needs of the core and to make them dependent on the export of a narrow range of primary products, e.g. nitrates in Chile, beef in Uruguay and Argentina, coffee in Brazil. The political consequences of this incorporation involved the maintenance and strengthening of the conservative power of landowners.

The age of imperialism

From the abolition of the Corn Laws in 1846 to the economic depression of 1873 Britain was the core power. British manufactured goods were sold throughout the globe, British capital was invested in most areas of the financial world and the economies of more and more peripheral areas were orientated towards Britain. The trading universe revolved around the 'economic sun of Britain'. The depression of 1873 marks a turning-point in the order of things. Other powers sought to construct tariff walls in order to protect their growing industries and to stimulate their own industrial base. Other countries began to catch up in the race to industrialize and the shifting of the centre of economic gravity in the core due to increased competition between the national economies is the essential ingredient of the *age of imperialism.* This age, which can be loosely dated as twenty years on either side of 1900, marks the extension of formal empires on a vast scale. The extent and pace of imperial expansion can be seen in Figure 1.2.

The age of imperialism has to be seen in terms of the rise of countries eager to achieve industrial growth and a place in the economic core of the world economy. The fusion of economic and political interests in the growth of neo-mercantilism is most clearly shown in the case of Germany. Welded together under Bismarck, Germany entered the last decades of the nineteenth century as a relative laggard in the race to industrialize. This was soon to change. A series of tariff measures protected German industry and by 1890 Germany surpassed Britain in the level of steel production. A strong school of neo-mercantilist thought began to emerge as the German Historical School was promoting protection for industry at home and the possession of colonies overseas. In all the major powers there was a growing body of opinion espousing the neo-mercantilist case based on

> the desire to build up a powerful national economy augmented by overseas possessions, whose production is geared to the needs of the mother country, while sharply delimiting similar rival systems and even isolating them from the entire home and colonial economy.
>
> (Gollwitzer, 1969, p. 63)

The force of this economic nationalism was soon felt in the international arena. In Africa, for example, European influence in 1880 was restricted to the Cape Colony, small areas of North Africa and some fringe coastal areas (Figure 1.3). Given the neo-mercantilist spirit, when Britain announced unilateral control of Egypt in 1882 and Germany claimed territory throughout Africa, the stage was set for large-scale partition. The atmosphere of claims, counter-claims, open hostility and rivalry was partly cleared by the Berlin–Africa conference called by Bismarck in 1884. The conference defined the conditions for future claims and although it achieved little by way of agreement over territory and spheres of influence 'by drawing up the rules of the game, it declared the game in progress' (Fieldhouse, 1966, p. 213). By 1890 the game was almost over and by 1914 the whole of Africa, apart from Liberia and Ethiopia, had been partitioned amongst the European powers (see Figure 1.4).

The economic forces stoking the engine of imperialism were given serious consideration by a number of contemporary writers. Two in particular deserve special attention because their writings have had such a profound effect on subsequent interpretations of imperialism.

Hobson and the tap-root of imperialism

Hobson's study of imperialism was first published in 1902, when the British empire was at its greatest extent. (The following quotes are taken from the 1938 third edition.) Hobson distinguished between colonialism and imperialism. Colonialism was defined as the migration of part of a nation's population to sparsely peopled foreign lands which

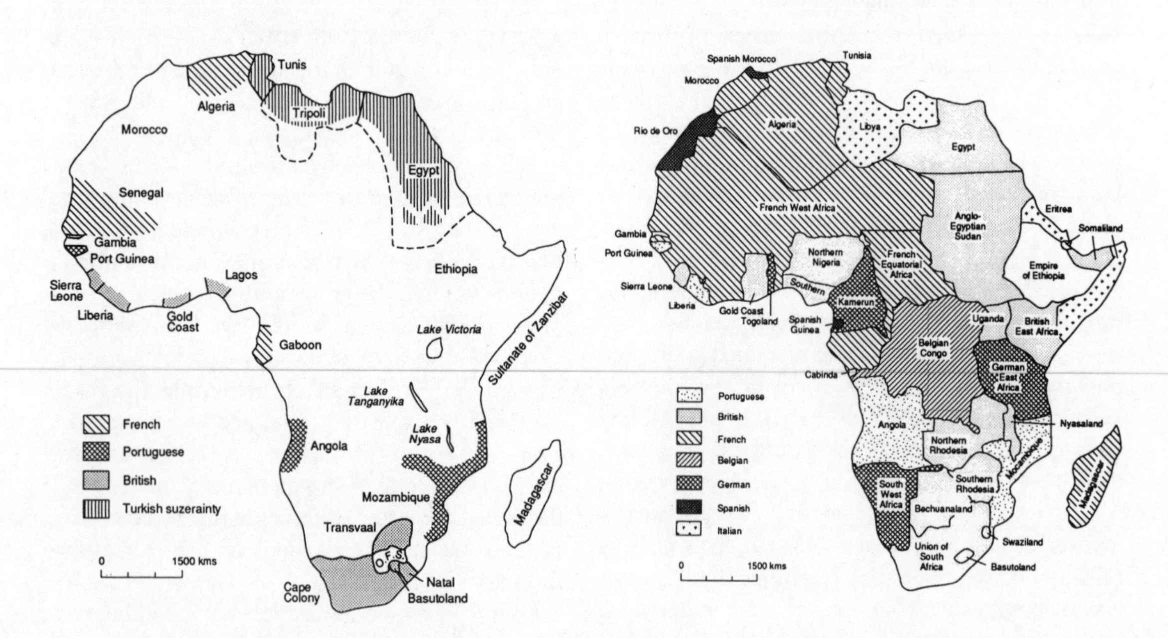

Figure 1.3 Colonialism in Africa, 1880

Figure 1.4 Colonialism in Africa, 1914

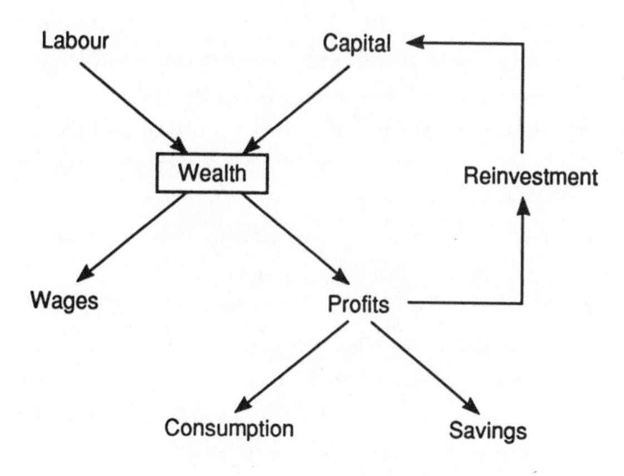

Figure 1.5 Flows of capital

eventually obtained some form of self-government. Colonialism, according to Hobson, was a genuine expression of nationality. Hobson had no quarrel with the colonial experience of Canada or Australia, though he seems to have given little consideration to the plight of the original inhabitants. Imperialism, on the other hand, was an artificial simulation of nationalism, involving costly expenditure and grave political risks.

Hobson argued that since 1870 Britain had obtained colonies, especially in Africa, which held very few commercial attractions, provided little room for emigration and gave only minimal trade. Why, given this background and the political risks involved, had Britain pursued this policy of imperialism? Hobson's reply was that certain sectional interests had guided foreign policy to their own ends. There were clearly identifiable interest groups which benefited from an aggressive imperial policy. Armaments manufacturers and army interests benefited from the increase in military and defence expenditure and shipping interests gained from the increased traffic, but the greatest gains accrued to capital investors. The most important economic impulse to imperialism was the influence of investors eager for Britain to take over foreign areas in order to secure new areas for profitable investment.

The economic drive to imperialism, according to Hobson, was the increase in the volume of production and capital accumulation on the one hand and the unequal distribution of income on the other. The increase in production meant an increase in wealth (see Figure 1.5). This wealth flowed in two channels. One flowed to labour in the form of wages and the other to capitalists in the form of profit, which went to the purchase of consumption goods, to reinvestment, or was held in the form of savings. Hobson argued that, in terms of this latter channel, there was a limit to the amount which could be spent by capitalists on consumption goods and, because of the cheapness of labour, there was little need for large capital investment. The relatively low level of wages and hence the restricted purchasing power of workers also meant that there was little scope for capital investment in consumer goods production for the home market. The level of capitalist savings, therefore, grew and grew. There was underconsumption because of the unequal surplus capital seeking profitable investment. The economic tap-root of imperialism was

> the endeavour of the great controllers of industry to broaden the channel for the flow of this surplus wealth by seeking foreign markets and foreign investments to take off the goods and capital they cannot sell or use at home.
>
> (Hobson, 1938, p. 85).

Hobson not only described imperialism, he discussed possible methods of stopping it. He believed that it should be stopped because it raised the possibility of war while diverting energy away from social reform at home. Hobson's answer was to create a genuine democracy based upon the people's will, so that the state should not become involved in foreign policy adventures for the sake of a rich minority and that it should equalize the distribution of income so that surplus capital and underconsumption could not rise.

Lenin and imperialism as moribund capitalism

Hobson's arguments provided part of the basis for Lenin's analysis of the economic forces behind imperialism. In his pamphlet *Imperialism, the Highest Stage of Capitalism*, written in Zurich in 1916 and published in Petrograd in 1917, Lenin drew upon the work of Hobson and Hilferding. From Hobson, Lenin borrowed the timetable of imperial expansion, the notion of surplus capital and the fact that groups of investors influenced policy to secure profitable investment opportunities. From Hilferding's *Finance Capital*, published in German in 1910, Lenin took the concept of the increasing concentration of economic activity and banking into larger and larger combines producing a new form of capitalism, one in which the banks controlled the flow of investment to large-scale industrial concerns. Hilferding had termed this new capitalism 'finance capitalism'.

With the rise of finance capital comes the growth of surplus capital seeking profitable spheres for investment. Returns are highest in the periphery because capital is scarce, land prices and wages are low and raw materials are relatively cheap. The era of finance capital is thus marked by the export of capital to the periphery by the monopolies (oligopolies) which form international cartels, each competing against the others in the world market. The competition between these cartels (we would now call them multinationals) parallels the competition between nation–states. Indeed, the drive for colonial possessions emanates from these cartels, which seek to maintain their profitable enterprise.

> To the numerous 'old' motives of colonial policy, financial capital has added the struggle for the sources of raw materials, for the export of capital, for 'spheres of influence', i.e. for spheres for profitable deals, concessions, monopolists' profits ... and finally for economic territory in general.
>
> (Lenin, 1965 edition, pp. 149–50)

For Lenin, imperialism was defined as capitalism in that stage of development in which there was:

(a) concentration of production and capital in the hands of monopolies (oligopolies)
(b) the merging of bank capital with finance capital and the emergence of a final oligarchy
(c) the export of capital on a large scale
(d) the formation of multinational combines which divided and demarcated the world economy
(e) the territorial division of the world amongst the biggest capitalist powers
(f) the creation of a privileged section in the proletariat in the imperial nations and the growth of opportunism in the working-class movement, which militated against social revolution in the core. The revolutionary epicentre had shifted to the periphery.

The work of Hobson and Lenin has provided the base point for many subsequent studies of imperialism. Some of their work has been severely criticized. Both writers place too great an emphasis on underconsumption in the core economies. Subsequent history, however, has shown that overseas investment can increase while home consumption is increasing. Both writers also place too great an emphasis on the formal empires. The twentieth-century period of decolonization has demonstrated that explicit political domination of the periphery by the core is not a necessary condition for the survival of capitalism.

The analysis and predictions of Lenin and Hobson have, by and large, been overtaken by subsequent events. For contemporary analysis, however, they have left as legacy a number of important topics and key questions. In particular, they drew attention to:

- the export of capital
- the importance of overseas markets for core economies
- the power of multinationals in guiding foreign and domestic policy
- the crucial relationship linking economic interests and political conflict between core states.

They also raised the following questions:

- What is the form of the relationship between core and periphery?

- Who benefits from such relationships?
- What are the effects of such economic ties on the political relations between core countries, within core countries and between core and periphery?

We can return to our historical exposition by considering this last question in the context of the age of imperialism. The overseas expansion of the great powers had a number of implications for relationships between these powers. The rapid partition of the Far East and the scramble for Africa were part cause and part effect of the sharpening of conflict between the great powers. According to the neo-mercantilist doctrines the world's surface was finite, markets limited and, since overseas markets were essential, the colonial advance of one nation could be purchased only at the expense of another. This expression of national material interests caused the colonial collisions which studded the history of this period. Like great tectonic plates the spheres of influence of the imperial nations shunted together to cause violent eruptions at various points. The cataclysm was reached in 1914.

The relationship between the core and the periphery can be considered in terms of the role of the local elites and the anti-colonial reaction. The age of imperialism was one in which peripheral areas were incorporated into the world economy. Imperialism was the political function of the process of integrating areas into the economy. Robinson (1972) has argued that whether an area was incorporated into the formal or informal empire depended upon the character of the elites in the area. Collaborative elites were those which aided or facilitated European economic penetration, non-collaborative elites fought against such incorporation. Where there were strong collaborative elites, such as the landowners of Latin America, there was little need for direct political control by core states. But where the elites did not encourage incorporation into a European-dominated world economy, formal political control was necessary. The history of the British in India is one of imposing authority in the face of elite opposition. British expansion in India was backed by force and force continued to be the basis of British rule. The transition from informal to formal empire status was based on the need for the core state to uphold a collaborative system that was breaking down and to forestall the actions of other core states. As Kiernan (1974) wryly notes, you only need to intervene in a banana republic when the bananas fail to arrive.

The imposition of formal colonial power often provoked an anti-colonial response; in Africa, for example, French intervention provoked large-scale uprisings in Tunisia and the British faced rebellions in Egypt and Sudan. In the Philippines US involvement sparked off a costly war with nationalist forces which lasted from 1898 to 1902. Often such anti-colonial movements were weak, more often than not they were defeated, but by keeping the flame of resistance alive they provided the basis for the process which ultimately led to the collapse of formal imperialism.

Within the core countries imperialism was the 'moving force of the age' (Gollwitzer, 1969). Global politics assumed a central role in government policy and imperialism was an ideology which permeated the very pores of society. The character of internal politics was interwoven with the nature of foreign policies. Imperialism partially reduced social tension on the domestic scene by promoting a unifying theme of national chauvinism. Imperialism as a unifying force was particularly strong in Germany, where the state used foreign policies to cope with potential domestic crises caused by rapid economic growth (see Hehler, 1972). Such policies were not new; according to Shakespeare, Henry IV counselled his son to

> Be it thy course to busy giddy minds
> With foreign quarrels.
>
> (Henry IV, Part 2)

In the age of imperialism, however, aggressive foreign policies to secure economic gains were used to counter the rise of working-class militancy. Cecil Rhodes notes, 'I have always maintained that the British Empire is a matter of bread and butter.

If you wish to avoid civil war then you must become an imperialist' (quoted in Gollwitzer, 1969, p. 136). In the UK most of the imperialist pressure groups tied their foreign policy objectives to definite national goals which would vitiate the potential for social unrest. The Liberal imperialists were strong advocates for social reform and the Tariff Reform League emphasized the full employment opportunities that would follow from the abandonment of free trade. Lenin's point was that imperialism had been too successful in these respects. He argued that the imperialist policies of the core countries, especially Britain, had created a privileged section of the working class. Since this labour aristocracy had a stake in the existing state of affairs he thought it unlikely that they would adopt a revolutionary posture. The wealth of imperialism had blunted the revolutionary edge of the working class in the core but the process of imperialism had raised the possibility of revolution in the periphery.

Geography and geo-politics

The effects of the 'moving force of the age' can be seen in the development of geography as an established discipline. The number of geographical societies grew enormously in Europe over this forty-year period. Overlying natural curiosity in the external world, a curiosity increasingly whetted by improvements in transport, was now placed the demands of national security, national wealth and international prestige. Geography was, as Mackinder had it, 'enlisted as an aid to statecraft and strategy'. The close connection between geographical research and national interest can be seen in the society formed in Germany in 1878 under the title 'Society for Commercial Geography and Promotion of German Interests Abroad' and in the tenor of contemporary geographical publications. Younghusband's (1910) book, for example, published under the title *India and Tibet* gives a moral justification for the British invasion of Tibet in 1904.

Perhaps the most famous geographer of the age was Halford Mackinder and his celebrated paper

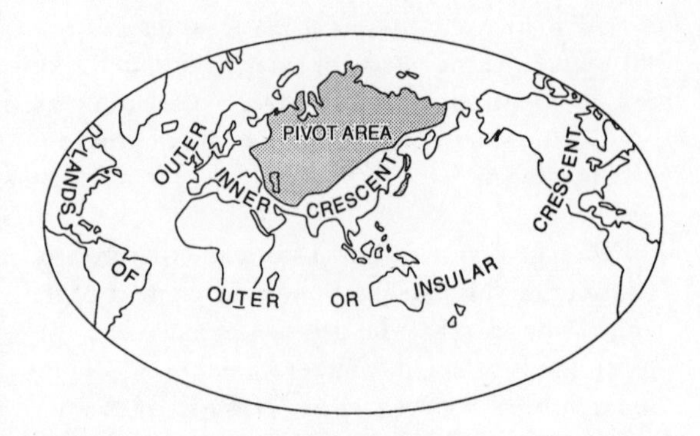

Figure 1.6 Mackinder's world view

'The geographical pivot of history' encapsulated the relations between national interests and geography. This paper was first read to the Royal Geographical Society in London in 1904. It is important to remember the context. Mackinder was writing at a time when various groupings of imperialist powers were emerging. Two years before Mackinder read his paper Britain had signed a treaty with Japan in order, as the two countries saw it, to deter Russian expansion; two weeks after the RGS meeting the Russo-Japanese war broke out.

Mackinder's paper begins by drawing attention to the fact that the age of overseas expansion was over. The world now formed a closed political system, the imperial powers were facing in on each other and political activity on the part of any one power would find an echo and a response in the others. On the basis of cursory historical evidence Mackinder then argued that the pivot region of world politics was the huge area of Euro-Asia (see Figure 1.6). This core area, the *heartland*, occupied a key strategic position in the world alignment of powers because it pressed upon the borders of many countries. It was now controlled by the Russians, who could develop the economic potential of the region free from any oceanic (British) influence. Around this pivot area Mackinder delimited an *inner crescent of marginal states* and an *outer crescent of oceanic powers* including Britain, the USA and Japan. Mackinder drew two strategic conclusions from his analysis:

- Russia (or whoever controlled the pivot area) should be prevented from expanding into the marginal lands because this would provide the basis for world domination. To prevent such an occurrence, signalled by the threat of a Russo-German alliance, a stable middle tier of independent states should be created between Russia and Germany.
- In the event of such a threat overseas powers should support armies in the bridgeheads of France, Italy, Egypt, India and Korea so as to force the pivotal powers to deploy land forces.

The British presence in India, for example, was therefore a necessary bulwark against the expansion of the pivot-area state into the south of the Euro-Asian land mass. In other words, Mackinder was giving a strategic rationale to the British presence overseas, a presence sanctified by the need to save world democracy and legitimated by the course of historical development in the post-Columbus era.

The close alliance of geography, geo-politics and national strategy was also found in the USA and Germany. In the USA, Bowman's (1922) book *New World: Problems in Political Geography* sketched out the implications of the 1919 peace settlement, as well as implicitly outlining an increased world role for the USA. Geo-politics reached its zenith (or maybe its nadir) in Germany, where Hausofer expanded Mackinder's work around the frame of German national interest. It was Hausofer who coined the term *lebensraum* to justify German expansion into Eastern Europe. The study of geo-politics has yet to recover from this experience; the damage to Eastern Europe was more serious. With the age of imperialism came the growth and strengthening of the world economy and the expansion of formal empires. The periphery was now used by the core economies as a source of raw materials, a market for manufactured goods and as a sphere for large-scale capital investment. The expansion of formal empires involved the extension of great power rule throughout the globe. An interdependent world economy and political system had been created.

THE CONTEMPORARY SCENE

Since 1945 core–periphery relations have been affected by a number of changes. Here we will examine two of the most important – *decolonization* and *economic imperialism.* The first suggests political independence, while the second indicates continued economic dependence.

Decolonization

Prior to the First World War anti-colonial feeling manifested itself in xenophobic outbursts against foreigners, but with some notable exceptions there were few stable mass movements and the small parties promoting national liberation were, by and large, formed around tiny groups of intellectuals.

Inter-war period

In the inter-war period there was a weakening of the imperial drive in the core countries. The immediate post-war era saw the growth of international policing and the Woodrow Wilson belief in

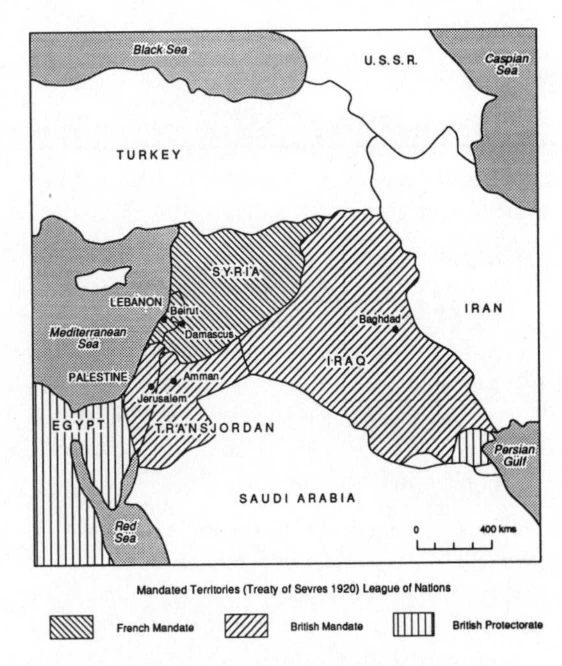

Figure 1.7 Middle East in 1930

Table 1.1 Distribution of mandates by the League of Nations

Mandate	*Territory*
A	Syria, Lebanon (France) Palestine (Great Britain) Transjordania (Great Britain) Iraq (Great Britain)
B	The Cameroons (France) North West Cameroons (Great Britain) Rwanda–Burundi (Belgium) Togo (France) Togo (West) (Great Britain) Tanganyika (Great Britain)
C	South West Africa (Union of South Africa) Territories of the Pacific – North of the Equator: the Caroline Islands; the Marianas Islands; the Marshal Islands (Japan) Territories of the Pacific – South of the Equator: Nauru (British Empire), administered by Australia Eastern New Guinea (Australia) Western Samoa (New Zealand)

the need for a new form of colonial relationship. Such beliefs were crystallized in the League of Nations mandate system for former German and Ottoman territories. Under this system the former colonies of the defeated powers were divided into a threefold categorization:

- Mandate A territories were ruled by either Britain or France, on behalf of the council of the League, with the intention that such territories would soon receive independence.
- Mandate B territories were thought unlikely to achieve independence except in the long term.
- Mandate C territories were thought incapable of being independent.

Table 1.1 notes the distribution of mandates throughout the world, while Figure 1.7 shows the pattern in the Middle East. The interesting thing about the mandate system is that it introduced the notion, new in international affairs, that former colonial territories could become independent, albeit after a period of time.

The inter-war period also saw the beginnings of much more vociferous anti-colonial feeling in Europe. The clarion call was sounded in the Soviet Union, where the Bolshevik Revolution had ushered an explicitly anti-colonial power onto the world stage. The First Congress of the Communist International in 1919 stated: 'Socialist Europe will come to the aid of liberated colonies with its technology, its organization, its spiritual forces in order to facilitate this transition to a planned and organized socialist economy'. At this time Bolshevik Russia had precious little technology, weakened spiritual force and a shattered organization; but the rhetoric was there and gave some hope to liberation movements and unsettled the colonial powers. The growth of criticism elsewhere in Europe varied from outright condemnation by Marxists to the more muted, measured tones of reprobation from such reformist groups as the British Labour Party.

In the peripheral areas of formal empire the growing pressure for some form of independence was strengthened by the economic depression of the 1930s. Deteriorating living conditions in the rural areas and mass discontent in the cities were given form and substance by the emerging intelligentsia and the growth of workers' movements. In India, Gandhi's first independence campaign began in 1920, the second in 1930. In Indonesia there was an attempted rebellion against the Dutch in 1926 which was primarily led by the local communist party. In Indo-China urban strikes and rural unrest broke out in the 1930s as the French sought to impose controls and checks on demands for independence. In Africa south of the Sahara, by contrast, things were relatively peaceful. There were few demands for independence since the policies of indirect rule had strengthened tribalism. There was thus little history of clearly demarcated nations; workers' movements or intellectual elites had yet to develop on any significant scale.

The post-war period

The turning-point in the retreat from empire came

with the Second World War. In the immediate post-war period the balance of forces was clearly swinging towards decolonization. The United Nations provided the forum for anti-colonial sentiments, and the Soviet Union and newly independent countries supplied the voices. The position of the USA, *the* world power, was more ambivalent. On the one hand the USA had an explicit anti-colonial stance and was eager to make friends and trading partners with the nations emerging from the struggle of the nationalist movements. But, on the other hand, US foreign policy after 1947 was dominated by the cold war and the aim of containing communism. Such a policy was based upon an alliance with the colonial powers of France and Britain, whose colonies provided valuable raw materials and strategic bases in the struggle against communism. To provoke too rapid a decolonization in these territories, US State Department officials argued, was to weaken the powers in the anti-Soviet alliance. The policy of the USA varied accordingly. In Indonesia the USA was hesitant about the anti-Dutch nationalists but did not intervene. In Indo-China, however, the anti-colonial struggle was seen through the ideological filters of the cold war. The US policy-makers argued that France had to win in Vietnam to withstand the threat of Russian tanks sweeping through the plains of Europe. It was a simple step, therefore, to see Vietnamese nationalist movements as a communist plot, pure and simple. When the French began to falter the US provided money and aid. By 1954 the US was paying 70 per cent of the French military budget in Indo-China.

In the periphery the process of decolonization occurred in two distinct phases in different parts of the world. In the first period, from 1945 to 1954, anti-colonial movements were strongest in Asia. The Japanese victories in the war had involved the destruction of European influence and given added impetus to all Asian nationalist movements. In British colonies there was a relatively peaceful transfer of power because the concept of the British empire had slowly been evolving, in response to nationalist pressures, towards the notion of a federation of states. British colonies were therefore bound, at least theoretically, for independence and the concept of empire entailed the prospect of independence. The process of decolonization was less peaceful in Indonesia and Indo-China. Both France and Holland wanted to return to the old relationship. The concept of the French Union, which the French established in 1954, entailed no prospect of independence. The Dutch in Indonesia were involved in quelling popular unrest in 1947 and 1948 until the strength of the liberation forces and the weight of international opinion resulted in independence in 1949. The tragedy of the nationalist movements in Indo-China was that their success in the early 1950s came at a time when the USA was seeing the periphery as the battleground on which to fight the communists. The liberation forces in Indo-China were dragged into the orbit of superpower intervention.

The decolonization in Asia strengthened anti-colonial forces in Africa where the second phase of the process began when Sudan achieved independence in 1956. Thereafter, independence for British colonies followed apace, each new state increasing the weight of precedence. Ghana obtained independence in 1957, Kenya in 1963, and Tanzania in 1964. Most of the French colonies achieved independence in 1960. The last colonial power left Africa in 1975 when the revolution in Portugal, partly caused by the cost of her colonial wars, meant overnight independence (after years of struggle) for Mozambique, Angola and Guinea-Bissau. It was another four years before Zimbabwe achieved independence and Namibia was controlled by the South African government until 1990.

Most of the sub-Saharan nationalist movements had no history of nationalism to rationalize their actions. The struggle for independence took place within the boundaries of colonial rule and the colonial boundaries were taken as the framework for new national boundaries. The present political map of Africa (Figure 1.8) bears a striking resemblance to the boundaries of the formal empires shown in Figure 1.4. The colonial legacy had its drawbacks. It perpetuated those states whose existence and boundaries were caused by the

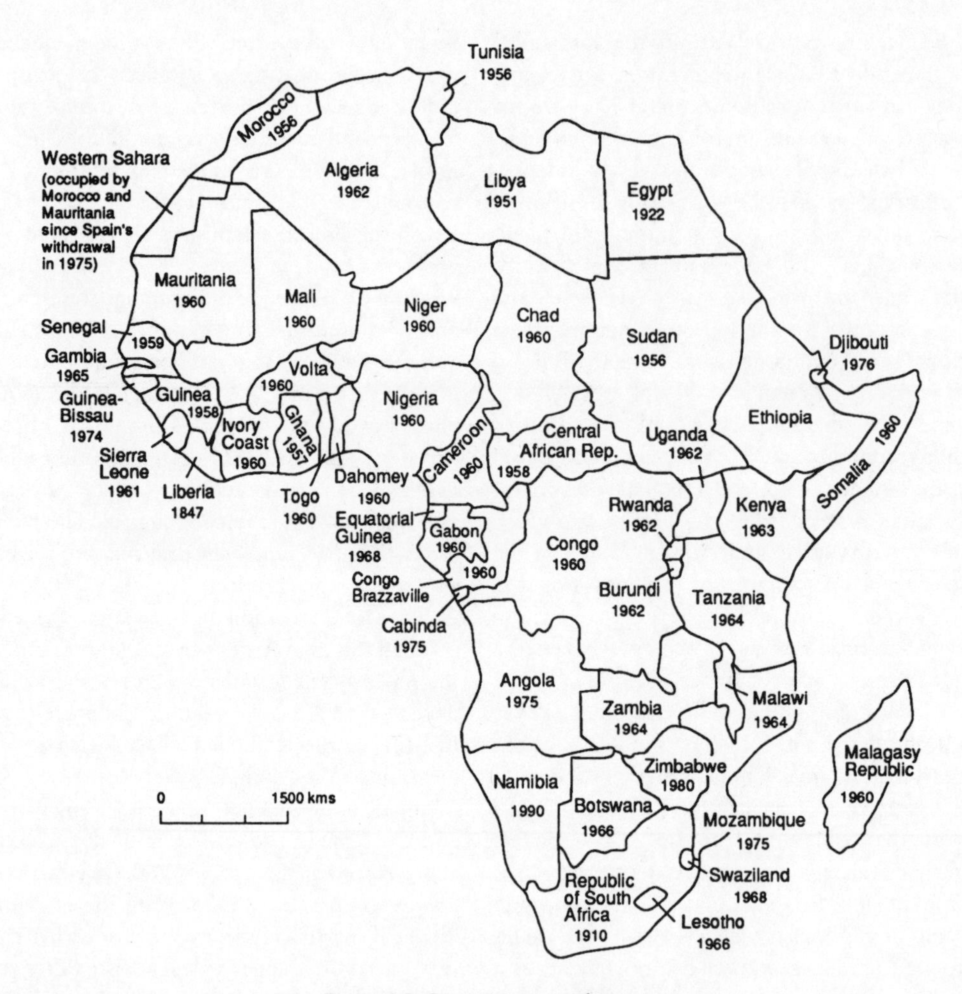

Figure 1.8 Contemporary Africa

machinations of competing colonial powers. Tiny Gambia in West Africa, for example, is 300 miles long and only 20 miles wide in places; its boundaries and orientation refer more to French and British rivalry in the nineteenth century than to contemporary social, cultural and economic realities. Similarly, the long sliver of the Caprivi Strip in Namibia, which touches Zambia and Botswana and Zimbabwe, owes more to nineteenth-century demands by Germany for access to the Zambezi river than anything else. The colonial boundaries also cut across tribal groupings throughout the continent. Many states were created which had an internal mosaic of tribal groupings or a small number of tribal groups, often locked into permanent conflict. After independence the unifying influence of anti-colonial liberation struggles could, and sometimes did, break down into tribalism. A major problem for politicians in Africa has been to weld different tribal groupings into nation–states. The motto of Zambia, 'One Zambia, one nation', is as much a plea as a slogan. Sometimes the tribal and ethnic differences can erupt into civil strife. Perhaps the most tragic to date was in Nigeria, where army officers of the Ibo tribe proclaimed the independent state of Biafra in eastern Nigeria in 1967. The bloody civil war lasted for three years. Nigeria had

BOX D: POST-WAR CORE–PERIPHERY RELATIONSHIPS

1 Decolonization has reduced the size of formal empires of core countries.
2 Decolonization has not been absolute. There are still fragments of empire, e.g. French possessions in the Pacific, Britain's continuing control over Hong Kong and the Falklands.
3 Core countries still exercise control:

- Directly, through agreements which are colonial in fact if not in name, e.g. the Trust Territory of the Pacific controlled by the USA.
- Indirectly, through economic relations, terms of trade, patterns of investment and financial relationships.

achieved political freedom, but like many other African nations, it remained a prisoner of the political geography of colonization.

Economic imperialism

Despite the political independence gained by peripheral countries, economic independence has been a more difficult goal to achieve. Political imperialism has been replaced by economic imperialism, which we can define as the effective economic control of the periphery by the core. Six aspects of this relationship will be considered:

1 The legacy of the past
2 Division of labour and terms of trade
3 Multinationals
4 Aid
5 Cartel power
6 Debt repayments

The legacy of the past

The history of colonialism and neo-imperialism imposed a definite structure on the world economy, one in which the economies of peripheral areas were orientated towards the needs of the core. At best the penetration of the periphery resulted in the growth of an export cash-crop economy and export-orientated enclaves which supplied raw materials to the core but did not engender local growth. The profits of such concerns were repatriated or used to buy foreign equipment, and local wages were used to buy foreign goods. The multiplier effects were exported to the core.

The growth of efficient cash-crop economies and limited industrial growth, against the background of large population growth, produced a large-scale marginalization of the masses. Unable to find a living on the land, they moved to the cities but many were unable to obtain jobs in the town. The economy and cities of many peripheral countries sagged and continue to sag under the weight of the underemployed and unemployed.

Frank's (1969) work on Latin America suggests that the most underdeveloped regions are those which had the longest and most sustained contact with the core. In similar vein Baran (1957) has contrasted the economic histories of India and Japan. India had been incorporated into the world economy in the nineteenth century when the process of de-industrialization and growth of cash crops led to India supplying the raw materials for, and buying the manufactured goods of, the British economy; in effect, the development of underdevelopment. Japan, by contrast, was not incorporated into the world economy. Owing to a complex combination of factors – the lack of natural resources, the subtle checks and balances on competing colonial powers and the nature of Japanese society – Japan remained relatively free

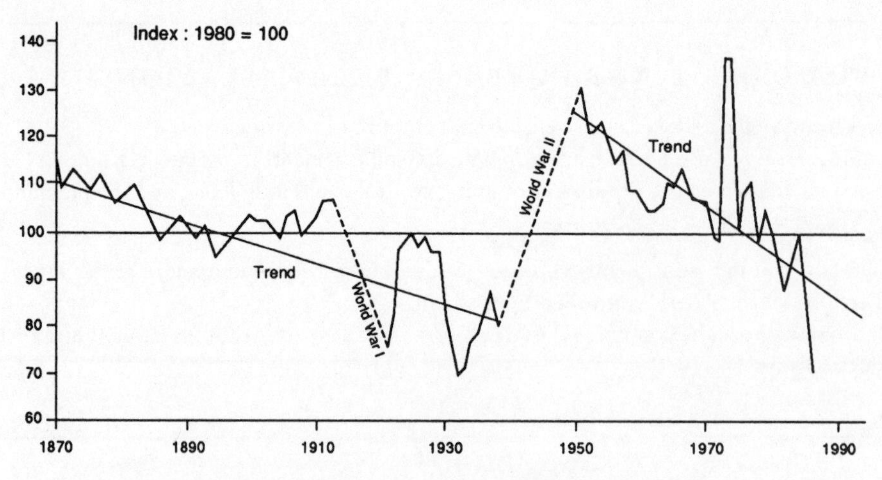

Figure 1.9 Rent commodity prices deflated by price of manufactures: 1870–1986
Source: IMF

from either US or European colonialism. It is this fact, according to Baran, that underpins subsequent Japanese economic growth and development.

Division of labour and terms of trade

The expansion of capitalism in the nineteenth and twentieth centuries spread the tentacles of trade throughout the world. The peripheral areas had a specific role to play in the international economic order, as their economies were orientated towards the export market to meet the needs of the core. Individual peripheral countries became dependent on a narrow range of primary commodities according to the pattern of climatic conditions and resource endowment. The economics of many peripheral countries became precariously balanced on the export performance of just one or two commodities. Even by 1986 90 per cent of Zambia's exports, for example, were in the form of copper. In Jamaica 60 per cent of exports were either bauxite or alumina, while 90 per cent of the export trade of Mauritius was sugar. Countries dependent on a narrow range of primary commodities are subject to periodic shocks caused by rapid fluctuations in commodity prices. Economies subject to such shocks are unlikely to experience real and sustained economic growth. Moreover, throughout most of this century there has been a widening disparity between the prices of primary and manufactured goods (see Figure 1.9).

The relationship between core and periphery has been elaborated (and sometimes mystified) by the concept of *unequal exchange.* Mandel (1962, Chapter 13) uses the term within the labour theory of value to argue that commodities from the periphery involve more labour than commodities produced in the core, due to the differences in labour productivity. The exchange of commodities between core and periphery at world prices thus involves a transfer of value from periphery to core. The form of the division of labour and the terms of trade involved in the world economic system has led to a transfer of wealth from periphery to core.

Multinationals

The threads which link periphery to core are spun by large multinational companies. These firms dominate the economic scene and the form of their activities maintains the core–periphery structure. Profits earned in the periphery can be siphoned off to shareholders in the core. This extraction can take the form of:

- repatriated declared profits
- royalty payments
- transfer pricing arrangements
- internal accounting mechanisms

The strategy of multinationals is based on criteria of profitability and market share rather than the development needs of the peripheral countries in which they operate. The chairman of a US-owned British subsidiary put the duties and loyalty of a multinational company executive very clearly:

> [He] must set aside any nationalistic attitudes and appreciate that in the last resort his loyalty must be to the shareholders of the parent company, and he must protect their interests even if it might appear that it is not perhaps in the national interest of the country in which he is operating. Apparent conflicts may occur in such matters as the transfer of funds at a period of national crisis, a transfer of production from one subsidiary to another, or a transfer of export business.
>
> (quoted in Tugendhat, 1973, p. 23)

The sentiments expressed by this executive take on potent meaning in peripheral countries where foreign-owned multinational companies dominate the economy. The economies of most peripheral countries are dominated by companies whose operations are guided by the dictates of profitability laid down in the core for the benefit of shareholders in the core. Such an arrangement need not necessarily lead to or contribute to the development of underdevelopment. In many cases it does.

Individual multinationals may be so powerful as to influence national policies to benefit their continued operations and profitability. Power is exercised in a number of ways. Indirectly, certain multinationals may seek to influence government policy by lobbying political decision-makers. The large companies will have considerable expertise in various forms of public relations exercises. More direct influence can also occur and the economic and political history of the peripheral world is littered with the effects of direct multinational company involvement on the course of political events. In 1910 the US-based United Fruit Company engineered an invasion of Honduras; in 1954 the same company in association with the CIA aided a rebellion in Guatemala which overthrew a government committed to land reform; in 1953 the major oil companies helped the overthrow of the Iranian government which was considering nationalizing the oil fields. More recently, the International Telephone and Telegraph Company (ITT) directly involved itself in the internal affairs of Chile. From 1970 to 1972 ITT was involved in both overt and clandestine attempts to prevent the election of Salvador Allende and its actions aided the military overthrow of his democratically elected government.

The ultimate expression of multinational involvement is non-involvement, when the state is essentially an adjunct of the corporation. The banana republics of Central America were the clearest expression of this relationship, and even today such arrangements can still be found. The recent political development of the Dominican Republic, for example, is to all intents and purposes the story of the state meeting the needs of one US company.

Subsidiary companies in peripheral countries seek to establish and maintain the availability of goods, raw materials and profits for their parent companies. Such aims could be achieved by promoting autocentric growth, income redistribution and a freer political climate. More often, however, the aims are achieved by backing the status quo. Recently, peripheral states have been demanding a greater say in the activities of multinationals. The threat, both actual and potential, of nationalization has involved companies in reaching planning agreements with peripheral states. This is an important trend but one which does not overturn the basic relationship of dependence. For that to happen, 'what will have changed is that investment, production processing, and prices will no longer be institutionally tied to the needs and strategies of global enterprises' (Girvan, 1976, p. 50).

Aid

An important strand in the core–periphery relationship is the aid contributions from core to periphery. In the immediate post-1945 period aid contributions were minimal. Both the USA and Europe were concerned with the reconstruction of the war-damaged European economies. Western interest in aid began as a response to the fear of Soviet influence in the periphery. The Twentieth Congress of the Communist Party of the Soviet Union in 1956 placed the USSR in a new posture with respect to the Third World. Emphasis was to be placed on aiding, at the least with ideological encouragement and at the most with money and guns, the revolutionary movements and nationalist governments in the periphery. The Western powers, especially the USA, countered the Soviet moves and began to use aid in the periphery to provide the basis for economic development which would maintain political stability and hence keep countries within the Western sphere of influence. The rise and fall of aid payments roughly corresponded with the high and low temperatures of the cold war.

The failure of aid to stimulate autocentric growth is also due in part to the form of aid. At best, aid has resulted in the construction of an infrastructure which promoted increased exploitation of a country's resources. Often this exploitation benefits the multinationals more than the local inhabitants. Aid for infrastructure merely maintains the economic relationship between the core and the periphery. At worst, aid can increase the dependency of the periphery on the core. Four aspects are important:

- The peripheral countries became burdened with debt service. From 1965 the rise in aid debt payments has meant that for most peripheral countries such payments are now greater than the capital received. The debt–service ratio of developing countries doubled from 1970 to 1980.
- Aid has been tied to the purchase of goods and services from the donor countries. US aid under the Alliance for Progress scheme in Latin America was distributed on the understanding that recipients would buy US goods, which were often more expensive than European and Japanese goods. Such restrictions and additional expenses led the president of Colombia to note that, 'Colombia has received two program loans under the Alliance. I don't know if we can survive a third' (quoted in LaFeber, 1976, p. 24).
- Aid has been used to maintain and reinforce the political links of dependence. Western aid has been given to discourage nationalization, to bolster governments friendly to the West, to stop the advance of left-wing parties and in general to maintain the status quo. Aid given by the USSR was similarly linked to political ends; it was given to maintain friendly socialist and non-aligned states.
- The organizations which allocate aid and development funds do so from the perspective of the core economies. The International Monetary Fund (IMF) and the World Bank, for example, are organizations set up by core countries to administer aid to peripheral countries in such a way as to maintain existing trading links and economic systems. The IMF in particular has come in for innumerable criticisms from peripheral countries for the kind of economic and social restrictions it places on receivers of aid. It has been argued that IMF conditions are so deflationary that they cause real hardship for the mass of the population.

Cartel power

We have noted in previous pages that many peripheral countries were, and still are, dependent on the export to the core of a few primary commodities. Against the background of worsening terms of trade many commodity-producing countries began to form cartels. The most important so far has been the oil cartel. The oil industry was dominated by the large multinational firms, the Seven Sisters as they were called, who determined rates of exploitation and the price of oil. When in 1959–60 the oil companies sought to decrease the price of oil from

$2.08 to $1.80 a barrel the effect on the oil producers was serious. Venezuela, a country where oil made up 90 per cent of exports, attempted to fight the price change. When this was successful Venezuela held talks with Middle East oil producers with the aim of providing a common front in negotiations with the oil companies in order to secure high prices. The result of these talks was the setting up of OPEC (Organization of Petroleum Exporting Countries) in 1960. The Original members were Iraq, Iran, Kuwait, Saudi Arabia and Venezuela. A year later Qatar joined and two years later Indonesia and Libya. During the 1960s Algeria, Ecuador, Gabon, Nigeria and Abu Dhabi joined the cartel. It had only limited success in its early days but as continued economic growth in the core sucked in raw material imports from the periphery the stage was set for the exercise of cartel power.

Oil was the life-blood of the economic boom of the early 1970s and OPEC began to bargain successfully with the oil companies. In a series of agreements OPEC managed to get the price of Persian Gulf oil (the benchmark of the oil trade) raised to $2.48 a barrel in 1972. Then, as inflation reduced the value of OPEC receipts, as demand continued to outface supply and as the fourth Arab–Israeli war broke out on Yom Kippur, OPEC used its cartel power with devastating effect. OPEC was dominated by Arab nations, who wanted to press home a political as well as an economic point. In December 1973 OPEC unilaterally raised the price of oil to $11.65 a barrel to take effect from January 1974. Later, OPEC members had control over pricing and production levels, and a second major price increase in 1979 took the price of oil to over $30 a barrel. The effects of OPEC actions have been global. Indeed, the economic history of the 1970s has quite simply been one of unfolding implications of OPEC actions. The hike in oil prices meant a redistribution of wealth. Formerly poor oil producers became some of the richest countries in the world. In the case of Iran the money was used by the Shah to carry out an extensive programme of modernization which carried the seeds of its own destruction. For the core countries the rise in oil prices had important effects but on the poorer peripheral countries the effects were severe. Unable to afford the increased prices, the poor oil-importing countries had to borrow on a high scale (see Table 1.2). By 1979 the international banks had lent $150 billion to oil-importing countries in the periphery.

The exercise of cartel power by OPEC signalled an important change in the structure of the world economy. The success of OPEC provided an example to other commodity producers. The number and composition of cartels formed before and after OPEC is shown in Table 1.3. In the short term these cartels can wield considerable muscle. In the long term, however, demand and supply are more elastic and producers have to fine-tune the price, otherwise their actions will encourage substitution and hence undermine their source of income. Commodity power waxes and wanes in response to the general economic climate; during boom periods imports are sucked into the core and the peripheral producers have considerable power, but during downturns and recessions the supply of commodities tends to exceed demand.

The success of a commodity cartel will depend upon sophisticated market knowledge, predictive ability and effective co-operation between producers. Cartels tend to be strong when just a few countries dominate production levels and weaker when a larger number of countries are involved. The greater the degree of similarity between the member countries, the greater the effective co-operation within the cartel. OPEC, for example, owes much of its strength to the shared Arab and Muslim experience. The strength of the cartel will

Table 1.2 Current account balances

	$US billion 1972	1977
OECD countries	8	9
OPEC countries	1.5	7
Non-oil producing countries of the periphery	−5	−36

Source: World bank

Table 1.3 Commodity cartels

Commodity	*Acronym*	*Date established*	*% of world exports*	*Countries involved*
Oil	OPEC	1960	76	See text
Cocoa	COPAL	1962	100	Togo, Nigeria, Cameroon, Ghana, Ivory Coast, Brazil
Copper	CIPEC	1967	56	Zaire, Zambia, Chile, Peru
Rubber	ANRPC	1970	98	Indonesia, Vietnam, Thailand, Malaysia, Sri Lanka
Tin	ITC	1971	85	Australia, Indonesia, Malaysia, Thailand, Zaire, Nigeria, Bolivia
Coffee	Coffee Mondial	1973	95	Indonesia, Brazil, Colombia, Angola
Bauxite	IBA	1974	80	Australia, Ghana, Guinea-Bissau, Jamaica, Haiti, Dominican Republic, Surinam
Bananas	UPEC	1974	92	Colombia, Ecuador, Panama, Costa Rica, Nicaragua, El Salvador, Guatemala, Honduras
Iron ore	AIOEC	1975	35	Australia, India, Sweden, Tunisia, Mauritania, Guinea-Bissau, Sierra Leone, Brazil, Venezuela, Peru, Chile

also depend on the nature of the commodity. Oil can be kept in the ground or conveniently stored for long periods of time. Bananas, by contrast, have to be sold quickly; the sellers have only limited power to withhold their produce. If they wait too long their crops, and hence their profits, are ruined.

Despite the problems of successfully wielding cartel power, producers' associations are now an important element of the contemporary economic and political scene. Their importance is likely to grow as natural resources become scarcer and as an increasing proportion of these scarce resources are located within the periphery. The lessons of OPEC have been learnt ... by both sides. After the oil price rises of the 1970s, for example, core countries sought to minimize their dependence on oil imports through substitution, energy conservation programmes and searching for oil deposits. The exploitation of the North Sea oil and gas resources was in part a function of oil price increases. A balance has been struck in the world economic order as primary producers seek to increase the price of their commodities. Importing countries, by contrast, seek to get these commodities as cheaply as possible. The resulting price is a function and a reflection of the balance of power.

Debt

An important element in contemporary core–periphery relationships is repayment of debt. In the 1970s and 1980s banks in the core lent money to many peripheral countries in order to finance development. The emphasis was on big prestige sites, car factories, petro-chemical complexes and the like. The conditions looked favourable – interest rates were low and, as Figure 1.9 shows, prices for primary goods were relatively high. Banks could see profits and the elites could see material gain.

The crunch came in the late 1970s in a mixture of interest rate increases which reached a crippling 21 per cent and a slump in the world economy. Many countries were stranded with high debt payments and reduced national income. The result was a debt crisis. The first indication of this crisis came in 1982 when Mexico announced that it could no longer pay its debts. This sent the banks into a panic: they had lent so much they had to carry on lending to enable Mexico to continue paying its debt. The scale of the debt crisis is enormous. In 1988 it was estimated that total Third World debt amounted to $1000 billion (see Table 1.4).

Table 1.4 The debt crisis

	Total debt ($ billion)	Debt repayments as % of export earnings	Debt per head of population
Argentina	59.6	55	1800
Brazil	120.1	40	840
Mexico	107.4	52	1346
Zambia*	5.8	10	820

*Zambia has reduced its debt service payment to 10% of export earnings, otherwise it would be paying over 60%.

Source: World Bank

In order to pay off these debts punitive fiscal measures were taken as governments reduced subsidies, and sought to decrease imports and increase exports. The result was a marked cut in the standards of living. In 1982, for example, wages in Mexico were halved and basic food prices were increased. The result was widespread malnutrition.

The debt burden was carried by the ordinary people of the Third World as austerity measures were introduced and standards of living were reduced. Susan George (1988) uses the term *financial low intensity conflict* to describe the erosion of living standards. The language is suitably militaristic: as one Brazilian commentator noted, the Third World War has already started, a war whose main weapon is interest payments.

The people who have to pay the debt are not those who benefited from the spending of borrowed money. Less than 10 per cent of the vast sum involved was used for productive investment. The rest was wasted in ill-conceived projects, graft and backhanders. The leakage of corruption was a flood which reached only a few very rich individuals. These elites are cushioned from the debt burden; the vast majority of the population did not benefit from the spending but have to suffer the costs of repayment.

POSTSCRIPT

The core–periphery concept is a simple model. Like all models it is an abstraction from the richness of reality. In reality, neither the core nor the periphery are homogeneous. We began this chapter by comparing selected figures for Bangladesh and the USA. Figure 1.10 shows how these are end points of a continuum through such countries as Saudi Arabia and Brazil. Figure 1.11 shows the more detailed subdivisions of the world as used by the World Bank. The differences within the periphery have been increasing. We can identify four distinct groups:

Oil-producing countries. Since 1974 the income of oil-producing countries has increased enormously. Vast oil revenues have flooded into what used to be relatively poor countries. Take the case of Saudi Arabia. In 1965 its economy was based on the export of a single commodity: oil. After the oil price rise of the 1970s this resource increased in value fourfold. This fed through to improvements in health. Life expectancy, which was about the same as in Bangladesh in 1965, improved while the population/physician figure tumbled to almost the same as that for the USA.

The oil producers are eager to diversify their economies. Dependence on a single commodity is a precarious position, as shown by the decline in the real price of oil in the 1980s. The oil producers have to widen their export base before the oil wells run dry if they are to constitute an area of autocentric growth outside of the core.

Middle income countries. These constitute a mixed bag and include Brazil, Mexico, Greece and the Philippines. These countries experienced accelerated rates of growth in the 1960s and 1970s. This growth was caused by:

(a) development of import substitution industries;
(b) increases in commodity prices;
(c) development of manufacturing industries.

In terms of (c) there has been what Peter Dickens (1986) refers to as a *global shift* in manufacturing employment, as capital investment flowed to those countries with cheap and docile labour forces. The

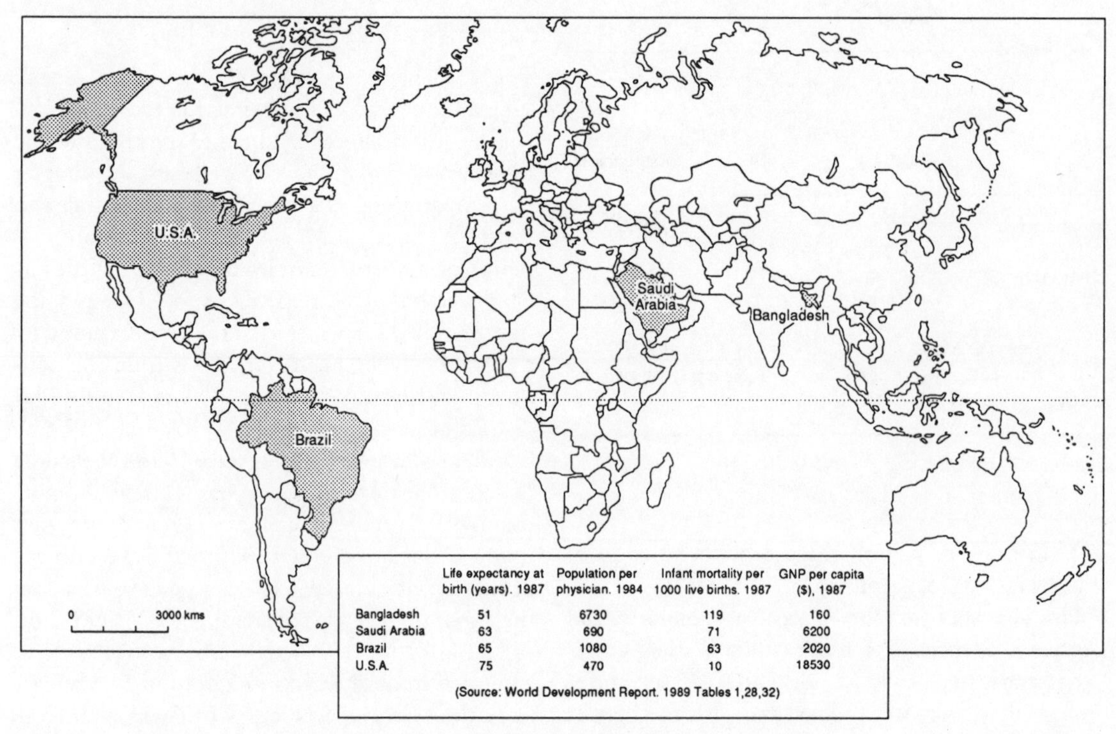

	Life expectancy at birth (years). 1987	Population per physician. 1984	Infant mortality per 1000 live births. 1987	GNP per capita ($), 1987
Bangladesh	51	6730	119	160
Saudi Arabia	63	690	71	6200
Brazil	65	1080	63	2020
U.S.A.	75	470	10	18530

(Source: World Development Report. 1989 Tables 1,28,32)

Figure 1.10 Selected statistics

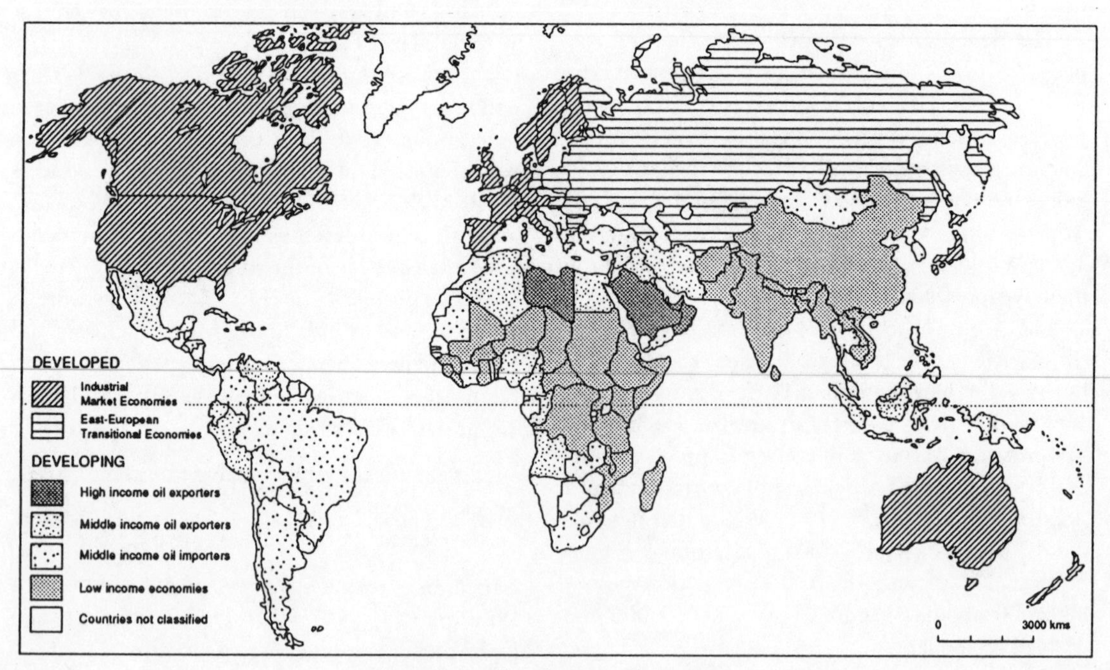

Figure 1.11 Classification of countries

BOX E: SUMMARY OF RECENT CHANGES IN WORLD ECONOMY

1 Low income countries have continued to remain poor. In relative terms their position has worsened and in a few cases, for example Ethiopia, their economic position has worsened in both relative and absolute terms.
2 Those peripheral countries with substantial oil deposits increased their export earnings. Dramatic rises in the 1970s have been followed by declining growth rates as the price of oil has fallen.
3 There has been a growth in manufacturing in selected countries of the periphery. We can now identify a subset of rapidly industrializing countries. Growth was greatest in the 1960s and 1970s. Recession and spiralling interest rates have placed an enormous debt burden on many of these middle-income countries.
4 In both relative and absolute terms the core countries have got richer.

middle-income industries provided a profitable source of investment. A new international division of labour is beginning to emerge in which increasing amounts of routine manufacturing processes are located in the middle-income bloc.

There have been some differences within this group. South Korea and Taiwan benefited from high influxes of capital as part of the USA's anti-communist strategies and Hong Kong and Singapore have developed as free-trade zones where the ideology and politics of growth and enterprise are strong. In Brazil and Mexico vast mineral resources have contributed to economic development.

For all these countries an important element in economic growth has been their ability to attract foreign capital because of low labour costs. Industrial growth has changed from import substitution and tariff-jumping industries to the development of industries selling goods on the world market. Multinationals have been attracted by cheap labour and governments which maintain order and allow relatively high levels of profit repatriation.

Growth has not been without its problems. The huge debt burdens of the 1970s and 1980s continue to plague many countries. The investment in industry by both private and state capital has made these countries very dependent on food imports and hence puts adequate nutrition on an unstable basis.

The benefits of economic growth have been unevenly distributed. Growth has been concentrated in the cities of selected regions. The process of uneven development has widened regional disparities and rural/urban differences in living standards. The resultant rural migration to selected urban centres swells the population of the cities and overwhelms the education and medical facilities. The process of rapid growth of capital-intensive industries against the background of agriculture shedding labour tends to produce the marginalization of a large proportion of the population. The under-employed and unemployed do not gain from investment in industry. This marginalization is reflected in a growing disparity between rich and poor.

In the 1980s growth began to falter. This was a result of crippling debt repayments, the emergence of even cheaper labour areas, a downturn in the world economy and new growth in the cores. As Figure 1.12 shows, growth rates in manufacturing fell relative to richer countries. The industrial growth of the 1970s was in the middle-income countries but in the 1980s growth took place in the richer countries.

Low income countries. Such countries include

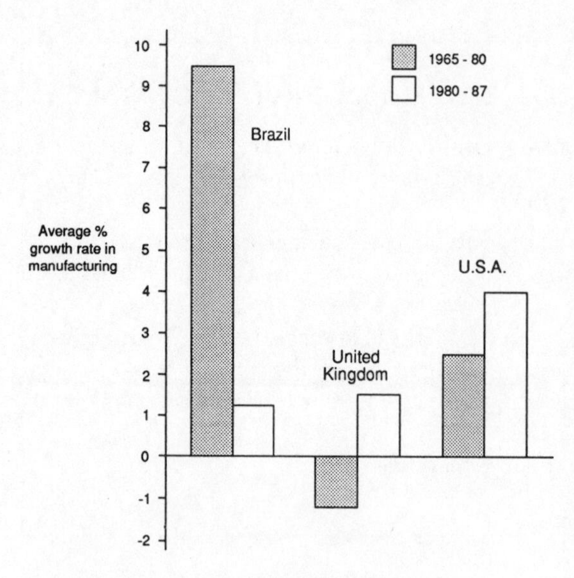

Figure 1.12 Growth rates in manufacturing
Source: World Development Report, 1989, Table 2

Bangladesh, Ethiopia, Mozambique, Indonesia and Haiti. These low income countries, thirty-seven in all, constitute 40 per cent of the world's population but have only 3 per cent of the world's wealth. Economic growth in these countries has been slow; their industrial base is slight and commodities few. Continued population increases and the price rise in imports have hamstrung strategies of economic growth. The disparities between low income countries, and the rest of the periphery and the core continue to grow and all the forecasts suggest that things will get worse. The numbers in poverty will increase, both absolutely and relatively, and the concept of the quality of life will be for the majority of people in these countries a meaningless abstraction. The position of these people constitutes a festering sore in the world body.

Transitional societies of Eastern Europe. The demise of state socialism in Eastern Europe has created a transitional category, as former communist countries move from state planning to some form of market economy. The situation is complex and constantly fluctuating as some countries break up, others are absorbed into larger entities, e.g. East Germany, and in others varying degrees of state control are maintained.

TOWARDS A NEW INTERNATIONAL ECONOMIC ORDER?

In the immediate post-war period the trading arrangements of the world were established by core countries in general and the USA in particular. When GATT (General Agreement on Tariffs and Trade) was set up in 1947, the development interests of the peripheral countries were not on the agenda and the peripheral countries were all but ignored. Things began to change, however, as peripheral states responded to continuing inequalities in wealth and deteriorating economic conditions. Because of the changing composition of the United Nations in the wake of decolonization this organization became a forum for peripheral opinion and international action. In December 1964 a UN resolution set up UNCTAD (United Nations Conference on Trade and Development). UNCTAD had three aims – to increase aid, to remove tariffs on the export of goods manufactured in the developing countries and to introduce commodity agreements which would protect peripheral countries from rapid and deleterious fluctuations in commodity prices. Despite its limited success UNCTAD was a pointer to the course of future North–South relations.

More recently, the relationship between core and periphery or North–South, as it is sometimes called, has been dominated by the periphery's demand for a new economic order. In 1974 the UN endorsed the establishment of a 'New International Economic Order'. The success of cartel power had given the old demands a new confidence. The new order was declared in the belief that:

> the present international economic order is in direct conflict with current development in international political and economic relations ... irreversible changes in the relationship of forces in the world necessitate the active, full and equal participation of the

developing countries in the formulation and application of all decisions that concern the international community.

(Quoted in Erb and Kallab, 1975, p. 186)

Similar sets of conclusions were reached by the Brandt Commission, whose report *North–South: a Programme for Survival*, published in 1980, also argued for changes in the old order. The report called for aid to be increased to 0.7 per cent of core countries' GDP, an international energy policy and large-scale investment in Third-World agriculture. The report argued from a position of enlightened self-interest. With under-utilization of resources in the core and poverty in the periphery it would seem sensible, the report suggested, to increase effective demand in the periphery.

Another report in 1987, *Our Common Future*, elevated the need for changes in core–periphery relations. The future course of core–periphery relations is likely to oscillate between the demands for a global New Deal and the demands for increased protection for jobs in the core countries. The demands for protection will be greatest in the weakest sectors of the core economies. The course of events will be decided by the balance of international economic power and national political power.

GUIDE TO FURTHER READING

Good historical studies of the evolution of the world economy include:

Barraclough, G. (ed) (1978) *The Times Atlas of World History*. Times Books, London.

Braudel, F. (1982) *Civilization and Capitalism 15th–18th Century, Vol 2. The Wheels of Commerce*. Collins, London.

Braudel, F. (1984) *Civilization and Capitalism 15th–18th Century, Vol 3. The Perspective of The World*. Collins, London.

Wallerstein, I. (1974) *The Modern World System*. Academic Press, New York.

Wallerstein, I. (1979) *The Capitalist World Economy*. Cambridge University Press, Cambridge.

Wallerstein, I. (1988) *The Modern World System. A second era of great expansion of the capitalist world economy 1730s – 1840s*. Academic Press, San Diego.

On the nineteenth century and the rise of Britain the work of Eric Hobsbawm is rarely bettered.

Hobsbawm, E. J. (1968) *Industry and Empire*. Weidenfeld & Nicolson, London.

Hobsbawm, E. J. (1975) *The Age of Capital 1848–1875*. Weidenfeld & Nicolson, London.

Hobsbawm, E. J. (1987) *The Age of Empire 1875–1914*. Weidenfeld & Nicolson, London.

The single most reliable book on imperialism is:

Gollwitzer, H. (1969) *Europe in The Age of Imperialism*. Thames and Hudson, London.

The death of formal empires is covered by:

Grimal, M. (1978) *Decolonization*. Routledge and Kegan Paul, London.

Various aspects of contemporary core–periphery relations are discussed by:

Alavi, M. and Shanin, T. (eds) (1982) *Introduction to the Sociology of Developing Societies*. Macmillan, London.

Brett, E. A. (1985) *The World Economy Since The War: The Politics of Uneven Development*. Macmillan, London.

Casson, M. (1986) *Multinationals and World Trade*. Allen & Unwin, London.

Crow, B. and Thomas, A. (1983) *Third World Atlas*. Open University Press, Milton Keynes.

Dickens, P. (1986) *Global Shift*. Harper and Row, London.

George, S. (1988) *A Fate Worse Than Debt*. Penguin, Harmondsworth.

Harris, N. (1987) *The End of The Third World*. Pelican, London.

Hayter, T. (1983) *The Creation of World Poverty*. Pluto Press, London.

Kidron, M. and Segal, R. (1984) *The New State of The World Atlas*. Pan, London.

Knox, P. and Agnew, J. (1989) *Geography of The World Economy*. Edward Arnold, London.

Singer, H. W. and Ansari, J. A. (1988) *Rich and Poor Countries. Consequences of International Disorder*. Unwin Hyman, London.

Thrift, N. (1986) 'The geography of international economic disorder', in R. J. Johnston and P. J. Taylor, (eds) *A World in Crisis?* Basil Blackwell, Oxford.

Programmes for reform are summarized in:

Brandt, W. (1980) *North–South: a Programme for Survival*. Pan, London.

World Commission on Environment and Development (1987) *Our Common Future.* Oxford University Press, Oxford.

Relevant journals

Development and Change
Economic Development and Cultural Change
Economist
Journal of Developing Areas
Journal of Development Economics
Journal of Development Studies
New Internationalist
South
Statistical Yearbook – published by United Nations each year since 1948, contains comparative statistics on trade, demography, health and welfare.
Studies in Comparative International Development
Third World Affair
Third World Quarterly
World Bank Atlas – published each year
World Development
World Development Report – published each year by the World Bank; contains a wealth of statistical data
World Economy

Other works cited in this chapter

Baran, P. (1957) *The Political Economy of Growth.* Monthly Review Press, New York.
Bowman, I. (1922) *The New World.* Problems in Political Geography. Harrap, London.
Erb, G. F. and Kallab, V. (1975) *Beyond Dependency: the Developing World Speaks Out.* Overseas Development Council, Washington, DC.
Fieldhouse, D. K. (1966) *The Colonial Empires.* Weidenfeld & Nicolson, London.
Frank. A. G. (1969) (2nd ed) *Capitalism and Underdevelopment in Latin America: Historical Studies of Chile and Brazil.* Monthly Review Press, New York.
Girvan, N. (1976) *Corporate Imperialism: Conflict and Expropriation.* Monthly Review Press, New York.
Hehler, H. U. (1972) Industrial growth and early German imperialism in R. Owen and B. Sutcliffe (eds) *Studies in the Theory of Imperialism.* Longman, London.
Hilferding, R. (1910) *Das Finanzkapital.* Vienna.
Hobson, J. A. (1938) (3rd ed) *Imperialism – a Study.* Allen & Unwin, London.
Kiernan, V. G. (1974) *Marxism and Imperialism.* Edward Arnold, London.
LaFeber, W. (1976) *America, Russia and The Cold War 1945–1957.* John Wiley, New York.
Lenin, V. I. (1965) *Imperialism, the Highest Stage of Capitalism.* Progress Publishers, Moscow.
Lichteim, G. (1974) *Imperialism.* Penguin, Harmondsworth.
Mandel, E. (1962) *Marxist Economic Theory.* Marlin, London.
Platt, D. C. M. (ed) (1977) *Business Imperialism 1840–1930.* Oxford University Press, Oxford.
Rees, J. F. (1929) 'Mercantilism and the colonies', in J. H. Rose, A. P. Nearton and E. A. Bervans (eds) *The Cambridge History of The British Empire,* Volume 1, Cambridge.
Robinson, R. (1972) 'Non-European Foundations of European Imperialism. Sketch for a theory of collaboration', in R. Owen and B. Sutcliffe (eds) *Studies in the Theory of Imperialism.* Longman, London.
Rude, G. (1972) *Europe in The Eighteenth Century: Aristocracy and the Bourgeois Challenge.* Weidenfeld & Nicolson, London.
Tugendhat, C. (1973) *The Multinationals.* Penguin, Harmondsworth.
Williams, G. (1966) *The Expansion of Europe in The Eighteenth Century: Overseas Rivalry, Discovery and Exploitation.* Blandford, London.

2

THE RISE AND FALL OF THE SUPERPOWERS: THE EAST–WEST FULCRUM

What is known as the 'Cold War' is the central human fracture, the absolute pole of power, the fulcrum upon which power turns, in the world. This is the field-of-force which engenders armies, diplomacies and ideologies, which imposes client relationships with lesser powers and exports arms and militarisms to the periphery.

(E. P. Thompson, 1980)

It is profoundly moving to see the forms of the old cold war declining before one's eyes, but declining most of all on the other side. The cold war has not been an heroic episode, an occasion for triumphs, but the most futile, wasteful, humanly destructive, no-through-road in history. It has led to conceivable investment in weapons with inconceivable destructive powers, which have – and which still do – threatened the very survival of the human species, and of other species perhaps more worthy of survival. It has nourished and reproduced reciprocal paranoias. It has enlarged authoritarian powers and the licence of overnight security survivors. It has abandoned imagination with a language of worst case analysis, and a definition of half the human race as an enemy other.

(E. P. Thompson, 1990)

INTRODUCTION

The second main dimension of the post-war world order is the dichotomy between East and West. The two blocs were headed by two countries, the USA and the USSR, and their interaction structured the post-war international political scene. World affairs from 1945 to 1990 were dominated by the USA and USSR.

In the following pages we will be examining the rise of these superpowers – a difficult task. Discussions regarding Soviet foreign policy were bedevilled by lack of information and massive inferences were made from the merest scrap of data or the slightest reshuffling of the party hierarchy. The Kremlin was, and is, not the most open of government centres and Winston Churchill's remark in 1939 still has some credence, 'I cannot forecast to you the actions of Russia. It is a riddle wrapped in a mystery, inside an enigma'. Moreover, the whole debate concerning the East–West split in world affairs has been contaminated by ideological axe-grinding, faulty judgement and just plain lies. Many analysts cut their analytical teeth during the cold war, with the resultant tendency to see the conflict as one of the good guys against the bad, freedom against tyranny, right against wrong. It is difficult to escape from this ideological fog which surrounds and envelops the truth in its stultifying embrace. But the attempt should be made.

THE USA

The roots of expansionism

The role that the USA played in the post-war world had its roots in earlier American history. From the very beginning the process of nation-building inevitably involved the USA in rivalry with the empires of Britain and Spain. By seeking to obtain more territory from other imperial powers the republic was stamped with the birthmark of expansion. In the north the USA sought to expand at the expense of the British empire and serious thought was given to 'liberating' Canada in 1812, but the US forces were militarily depleted by the Canadians. In the south and west, California, Texas and New Mexico were wrested from Mexico. In the drive to the west the native Indian populations were subjugated. In no other country, apart from the USSR, has expansion of frontiers played such a role. The effects have received conflicting interpretations. Turner (1963) associates democracy in the USA with the existence of an expanding frontier. The open lands of the west, according to Turner, provided a safety valve to the build-up of social pressure in the east: they ensured freedom and economic equality. Democracy and expansion went hand in hand. Democracy gave vitality to expansion and continued expansion ensured democracy. A belief in the Turner thesis has been one of the many strands which have made up post-war US foreign policy; it has often been argued that the USA's global role ensured democracy at home.

An alternative thesis has been suggested by W. A. Williams (1961), who argued that expansion has been a method for avoiding democracy and that it has diverted attention away from the creation of a truly democratic nation. Critics of US 1960s involvement in Vietnam, for example, adopted a position not dissimilar to Williams when they argued that the war deflected attention from social reform and civil rights.

By the early nineteenth century the USA was mapping out spheres of influence. In the Monroe Doctrine, enunciated in 1823, the USA propounded four principles for the conduct of international affairs:

1 the USA would not intervene in European affairs;
2 it would respect existing colonies of European powers;
3 recognized republics in South America should not be colonized;
4 any attempt to do so would be interpreted as an unfriendly act towards the USA.

The doctrine was partly defensive, partly expansionist. The USA was attempting to dissuade further European penetration in Latin America and, by telling the Europeans to keep clear, the way was to be left open for the exercise of US influence. Latin America was to be within the USA's sphere of influence.

The expansion of both the formal and the informal US empires only really developed after the Civil War. Although Walt Whitman could write in 1860:

> I chant the new empire
> I chant America the new mistress

this represents conceit rather than political reality. Prior to the Civil War there were limits to overseas expansion. The economy was only just beginning to realize its enormous potential and much time and energy was expended on westward expansion. The Civil War (1861–65) slowed expansion but after it ended momentum was regained.

Superimposed upon the forces of empire-building were the forces of industrialization in the economy and the drive of finance capitalism. Throughout the latter half of the nineteenth century there was a growing commercial impetus for expansion, shown in the consensus amongst businessmen, financiers and political leaders that further expansion was necessary to provide safe investment sites for their accumulating capital, possible markets for manufactured goods and as a method of preventing economic depression and social unrest at home.

In Asia the main focus of the USA's designs was China (see Figure 2.1). Alaska was acquired from Russia in 1867 partly as a back door to Canada, partly as a stepping-stone to Asia. In the same year Midway Island was annexed as a coaling port for US ships on their way to Asia. Pearl Harbour was annexed in 1887 for the same purpose and in 1893 the Hawaiian Islands as a whole were finally

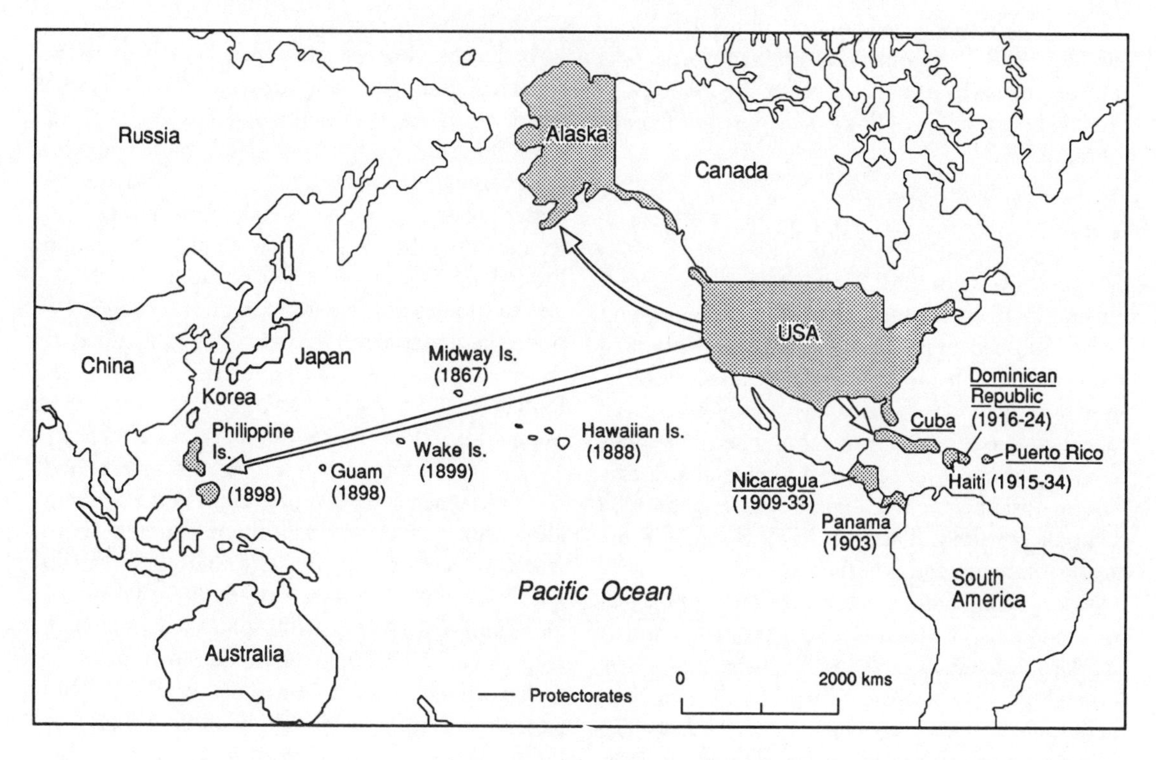

Figure 2.1 Early US expansionism

annexed. The Philippines and Guam were also obtained in 1898 as the victor's rewards in the Spanish–American war.

American business and finance lobbies were eager to get a share of the fabled hugeness of the Chinese market. Fearful of being eased out by the other imperial powers the US State Department in 1889 issued a series of Open Door notes which called for equal access to China. It was in China that the USA first came into conflict with Russia. The Russians were advancing eastward, the Americans westward. The designs of the USA on Manchuria were thwarted by the Russo-Japanese treaties of 1907 and 1910 which divided Manchuria between the two signatories. The roots of the conflict between the two superpowers were thus put down well before the Bolshevik Revolution.

There was also US expansionism into the Caribbean after Spain's defeat in the Spanish–American War. Puerto Rico was annexed at the same time as the Philippines; Cuba was not made part of the formal empire but under the Platt Amendment of 1901 it had restrictions placed on its foreign relations and the size of the national debt. It also had to support US actions and provide military and naval bases for the US armed forces. Despite this supposed political independence the USA directly intervened in 1906, 1911 and 1917 to quell domestic upheavals which threatened US business interests. The USA was active elsewhere in the Caribbean. Punitive expeditions were sent to Mexico in 1911 and 1916, and a number of other countries experienced indirect involvement in the form of being declared protectorate states under US guidance (see Figure 2.1).

By the time the nineteenth century turned into the twentieth the infrastructure of expansion had been laid. Overseas possessions had been gained and an economy was developing which was dependent on overseas trading and financial links; underlying the economic and political forces was an ideology which in the name of free enterprise and democracy was willing to sanction interference abroad.

The rise to globalism

The end of the Second World War saw the USA emerge as the richest and most powerful nation on earth. Subsequent policy was based on maintaining this position. The most important characteristic of post-war US foreign policy has been the essential oneness of economic, political and security interests. Economic goals, political aims and strategic considerations were fused into an essential unity in US policy-making. The USA has a key role in the world economy and is intimately associated with overseas economies, which it needs for supplies of raw materials, as sites for capital investment and as markets for agricultural and industrial produce. Foreign-policy objectives are primarily aimed at maintaining and servicing these links. The symmetry of economic, political and military objectives was perhaps best described by a former treasurer to the General Electric Company. 'Thus our search for profits places us squarely in line with the national policy of stepping up international trade as a means of strengthening the free world in the Cold War confrontation with Communism' (quoted in Magdoff, 1969).

The early years

As the Second World War drew to a close the USA attempted to secure an Open Door policy. To this end the Bretton Woods conference in 1944 created free-trading arrangements (allowing the USA access to British home and colonial markets) and the international organizations of the International Monetary Fund (IMF) and the World Bank. Since voting rights on these organizations' committees were in proportion to financial contribution and because the USA was the largest single contributor, it effectively controlled the institutions. Post-war economic growth was to be financed and controlled by the USA.

In Eastern Europe the Open Door was to be slammed shut. The Red Army had marched from Russia to the gates of Berlin and was in control of most of Eastern Europe. The USSR wanted governments friendly to it installed to act as a defensive buffer against possible attack. The USSR also wanted the resources and machinery of Eastern Europe to replace its shattered economy. The USA, on the other hand, wanted the door left open for trade and investment. When the USSR extended its sphere of influence in the region the scene was set for conflict. From the Soviet perspective it was essential to hold Eastern Europe as a defence against attack and as a reservoir of resources and transportable machinery: the claims of the Western allies ignored the world position that the USSR was entitled to after its role in the war, a war in which 20 million Soviet citizens died. From the viewpoint of the Western allies, Soviet actions were all preludes to attempted world domination.

In such a climate of mutual distrust, battle lines were drawn up. In 1946 Churchill, in a speech at Fulton, Missouri, noted that 'from Stettin in the Baltic to Trieste in the Adriatic, an iron curtain has descended across the continent' (see Figure 2.2). According to the State Department in Washington, uprisings in Greece and Turkey were inspired by the USSR, whose ultimate aim was world conquest. Greece and Turkey were considered front-line states; if they fell, then, like a set of upright dominoes which have been pushed over, Europe would also fall. This was the beginning of the domino-theory motif in US foreign policy. It was a policy that, as Table 2.1 demonstrates, was infinitely adaptable across space and time.

The years of 'irreconcilable conflict'

The response to the perceived Soviet threat was twofold. First, the Truman Doctrine, announced in March 1947, stated that it was the policy of the USA to support 'free peoples who were resisting attempted subjugation by armed minorities or by outside pressure'. The term 'free' was never defined, but was interpreted in terms of openness

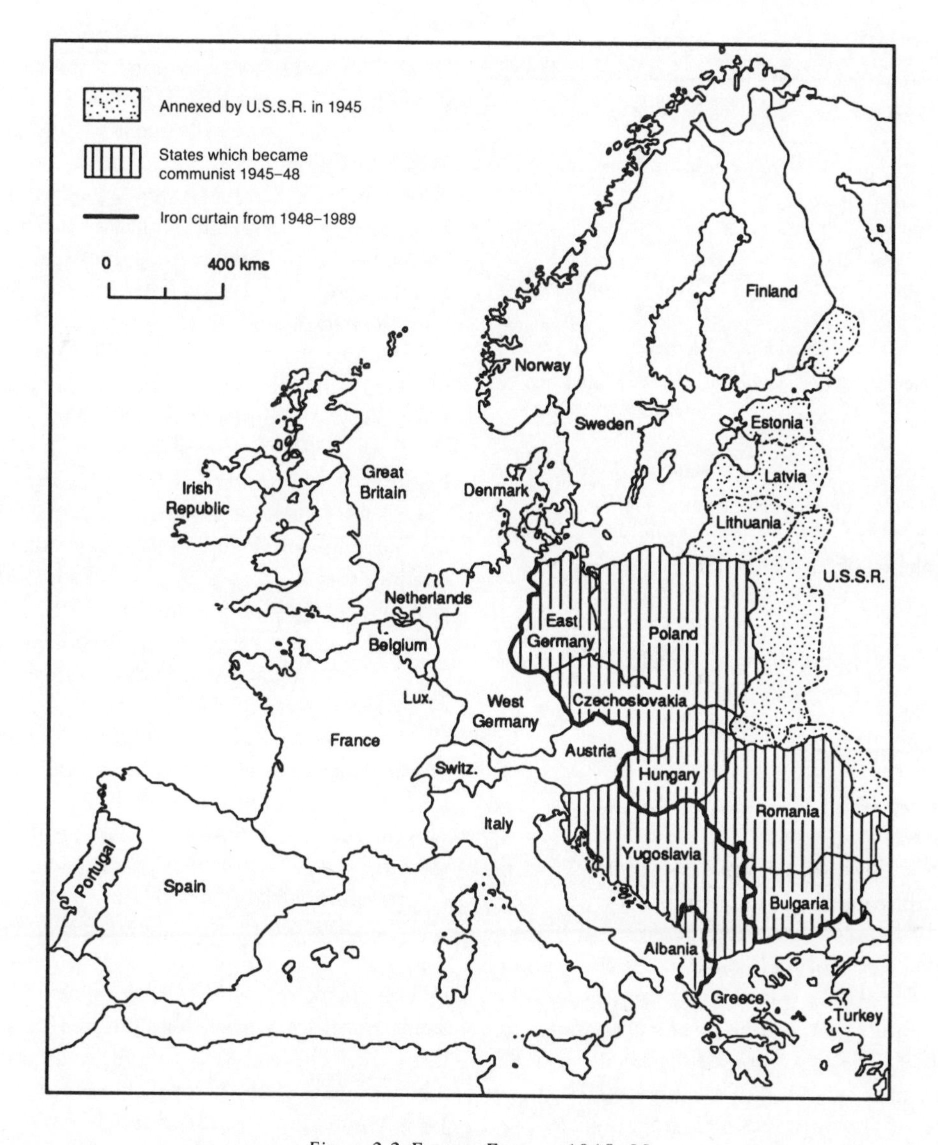

Figure 2.2 Eastern Europe, 1945–89

to US business interests and receptivity to US strategic considerations. The doctrine's promise became a useful fall-back for any dictator facing social unrest; the blame could always be placed on the influence of communist organizers. Any government in a sensitive area could use the communist bogey to get US aid to bolster its power and authority. Conceived as a tract for freedom, the doctrine became a blueprint for repression.

The second response was the Marshall Plan, unveiled in June 1947, which involved $17 billion aid to Western Europe. The money was to generate economic growth so as to counter the influence of communist and socialist parties; Western Europe would then be a stronger bulwark against the Soviet Union and its economic development would boost the US economy. In the Marshall Plan, as in other US policies, economic considerations were

Table 2.1 The domino theory in US foreign policy

Date	*Speaker*	*Dominoes*
1947	Acheson (Under-Secretary of State)	Greece → Turkey → Europe
1948	General Clay (Army Commander in Germany)	Berlin → West Germany → Europe
1954	Eisenhower (President)	Vietnam → S.E. Asia
1955	Eisenhower (President)	Quemoy/Matsu → Formosa → West Pacific
1961	Kennedy (President)	Berlin → Europe
1964	Rostow (presidential adviser)	South Vietnam → Thailand → S.E. Asia
1964	Johnson (President)	South Vietnam → Hawaii → San Francisco
1980	Carter (President)	Afghanistan → Iran → Gulf States
1990	Bush (President)	Kuwait → Saudi Arabia

interwoven with political goals and strategic objectives.

Elaborations to the initial Truman Doctrine were to provide the basis for post-war US foreign policy. The policy was essentially one of building up armed forces of both nuclear and conventional types. The USA was to act as the world's policeman in the defence of 'democracy' and 'free enterprise'. The expansion of the USSR would be stopped in its tracks and US business would be allowed to prosper. Such a policy involved a system of:

1 alliances
2 containment of Soviet 'advances'
3 interventions: direct, small-scale, military and economic involvement where appropriate.

Alliances: In 1938 the USA had no military alliances and no troops stationed on foreign soil. In 1959 the USA had 1400 foreign bases in 31 countries, and by 1989 military alliances had been signed with 50 states and just over 1.5 million service personnel were stationed across 117 countries. The three most important alliances were NATO, SEATO and CENTO, which tied the USA and its allies to involvement in, respectively, Western Europe, South-East Asia and Central Asia. The NATO treaty was signed in 1949 by Belgium, Canada, Denmark, France, Iceland, Italy, Luxembourg, the Netherlands, Norway, Portugal, Britain and the USA. Later signatories were Greece, Turkey and West Germany. The SEATO organization lasted from 1945 to 1975 and included Australia, Britain, France, New Zealand, Pakistan, the Philippines and the USA. Signed in 1955, the CENTO treaty involved Britain, Iran, Pakistan, Turkey and the USA. From the US viewpoint such alliances bound together the countries on the edge of a supposedly expanding Soviet empire; they were the mortar which bound the crumbling edges of an unstable edifice. For the USSR the alliances were interpreted as a policy of encirclement, a giant noose drawn around the neck of the Soviet state.

Containment: The USA sought to halt perceived Soviet expansion by a policy of containment. The USA was to provide countervailing pressure wherever in the world Soviet-inspired revolution was seen to exist. This policy of containment led to US troops fighting in Korea and South Vietnam.

When the North Koreans attacked the South across the 38th parallel on 25 June 1950 it was quickly perceived in Washington as an expansionist move of world communism. The invasion threatened the stability and the markets of the whole of Asia. The USA was quick to act. On 27 June it managed to push through a resolution in the UN calling for assistance to South Korea and by September 1950 troops led by General MacArthur had landed at Inchon. The battle lines swung backwards and forwards. The UN forces, 90 per cent of whom were from the USA, reached the Chinese border by October 1950 but by January 1951 a combined Chinese and North Korean force had pushed the front back into South Korea. The war dragged on until 1953, when an armistice was signed. The peace treaty kept the border at the 38th parallel. Approximately 5 million people had died by the end of the war; the frontier remained the same as at the beginning.

In its desire to end the war quickly the USA used the threat of massive retaliation. Eisenhower had threatened the Chinese with nuclear weapons if the treaty was not signed quickly. Containment could now be achieved by virtue of nuclear superiority. The problem with massive retaliation, however, was that it was an all-or-nothing response, useful to counter the great threat but useless against small-scale liberation movements.

The policy of containment in the context of liberation movements reached its culmination in Vietnam. After the French defeat at Dien Bien Phu in 1954 the USA shouldered more of the burden of fighting, as part of the policy of containment. Officials agreed that it was important to show the communists, especially the Chinese, that liberation movements would not succeed. The Third World had to be shown that the USA would stand up against perceived communist incursions. Containment had to be pursued, it was argued, for if Vietnam fell it would herald the fall of South-East Asia to the communists. Other dominoes would keel over. Important markets and vital raw-material supplies would thus be withdrawn from US economic influence. The USA fought in Vietnam to preserve its economic might, to bolster its image as the world's policeman and to impose its world view on the unfolding of events in the periphery. Even a brief discussion of the events makes salutary reading.

The Geneva Convention of 1954 established the division of Vietnam into North and South along the 17th parallel. The convention also stated that elections should be held. They never were in the South because the USA feared that Ho Chi Minh would win – and this from the upholder of democracy and freedom. From 1955 onwards the USA supported Ngo Dinh Diem and provided military aid. Diem's regime was very unpopular; it was seen to be corrupt and repressive and Diem was a rich Catholic in a country where three-quarters of the population were poor Buddhists. Resistance to Diem hardened and in 1960 delegates from a number of nationalist groups formed the National Liberation Front (NFL) with the aim of replacing Diem. In the post-war context Diem could raise the spectre of communist insurgents to receive yet more aid. The scene was set. The USA was, as it saw it, trying to defeat communism in south Vietnam, but by labelling critics of the Diem regime as communists it did more to strengthen the hand of the communists in the NFL than anything else. All critics of the corrupt regime, of whatever affiliation, became by definition communists or communist sympathizers.

The war escalated during the 1960s. In 1960 there were fewer than 800 US 'advisers' in Vietnam. By 1963 this had risen to 17 000 troops, by early 1965 to almost 170 000, until in 1969 555 000 US troops were stationed in Vietnam. Because the strategists believed that the liberation movement was a communist plot masterminded from the North and since the NFL was receiving support from there, US strategy focused on the North. President Johnson deceived the US Congress about the supposed Gulf of Tonkin incident and in the subsequent resolution of the same name the President was given a blank cheque to pursue military ends. The cheque was cashed in with the saturation bombing of the North. By 1969 over seventy tons of bombs had been dropped for every square mile of North Vietnam. As the war escalated, so did the casualties. During the mid- to late-1960s, US casualties rose to approximately 250 a week and the full horrors of the war were being seen in most sitting-rooms, courtesy of television. As the casualties increased and the concept of an American military victory faded from sight, resistance to the war began to grow in the USA. As the war proceeded further to suck in men, money and material in a maelstrom of death and destruction, protest spread from the students and intellectuals to the rich and powerful. In 1968 peace talks began in Paris and Nixon became president with vague murmurings of putting an end to the war.

The winning of the peace proved as bloody as the attempt to win the war. Nixon's policy of Vietnamization involved reducing the number of US land troops by increasing aid to the South and by providing more US air power. In order to destroy the supply routes from the North, US and South Vietnamese troops invaded Cambodia in 1970 and the US Air Force provided cover for the South Vietnamese invasion of Laos in 1971. Criticism of the war continued and by 1973 Congress had prohibited the reintroduction of US forces in South Vietnam. As the USA pulled out and the North Vietnamese advanced, the regime in the South collapsed. By 1975 Vietnam, Laos and Cambodia all had communist governments. The policy had failed; it had cost the lives of 46 000 US service personnel, 2 million Asian troops and 1.5 million civilians.

Small-scale intervention. Vietnam was a failure. The USA had paid a high price in its unsuccessful bid to contain a national liberation movement. Elsewhere, the USA had achieved economic and military objectives by a series of successful small-scale operations. The following are only a tiny proportion of US activities.

In Latin America, the Monroe Doctrine was interpreted as a justification for US intervention. In 1953, for example, the President of Guatemala confiscated the property of the United Fruit Company. In 1954 the USA airlifted arms to rebels in nearby Nicaragua and Honduras. The CIA and the US embassy in Guatemala aided the coup which installed army officers acquiescent to the demands of the United Fruit Company. In the Dominican Republic the US government gave support to a *coup d'état* in 1962 which replaced the radical president Juan Bosch by the conservative Reid Cabral. When unrest against the unpopular Cabral broke out in 1965 order was restored by the presence of 23 000 US marines who landed in the capital of Santo Domingo. The very success of the operation gave further support to those in the Pentagon and the State Department who argued for greater resolve in containing communism in South Vietnam and for more direct US involvement in South-East Asia. In 1983 US marines landed in Grenada and in December 1989 US troops invaded Panama in the most recent example of the vitality of the Monroe Doctrine.

The US government also intervened in the Middle East. In 1953 an Iranian nationalist movement headed by Mohammed Mossadegh confiscated the holdings of the Anglo-Iranian Oil Company. The move was widely supported within Iran, where in the same year Mossadegh had received over 90 per cent in a plebiscite. The US government was less pleased. All aid to Iran was cut and arms, equipment and money given to the Shah to help in his bid to regain power. The bid was successful. In the ensuing discussions the US oil companies received 40 per cent of Iran's oil production, to which the USA had previously been denied access by the British-owned Anglo-Iranian Company. More direct US involvement took place in Lebanon. In 1958 the US government feared restrictions on Middle-East oil supplies. Nasser's brand of Arab nationalism was gaining support and in the Lebanon civil war was being waged between Maronite Christians and pro-Nasser Moslems. When the Christian president asked for support the USA landed 14 000 troops on Lebanese beaches. Dulles explained to Congress, 'It was time to bring a halt to the deterioration in our position in the Middle East.'

The recorded interventions are only the tip of the iceberg. Submerged beneath the waves of secrecy lie the endeavours of the CIA to bolster friendly governments, topple unfriendly ones, get rid of troublesome politicians and advance the cause of US interests.

Reiteration of the Truman Doctrine

The Truman Doctrine has been the corner-stone of US foreign policy since it was announced in 1947. In various forms, at different times and in changing circumstances, it has been reiterated by successive presidents. In 1957 the Eisenhower Doctrine promised US military and economic aid to any nation in the Middle East which requested help against communist-inspired aggression. Kennedy's position was eloquently stated in his 1961 inaugural address, '... we shall pay any price, bear any burden, meet any hardship, support any friend, oppose any foe, in order to assure the survival and success of liberty. This much we pledge – and more.' In justifying the landing of marines in Santo Domingo the Johnson Doctrine of 1965 announced, 'American nations cannot, must not and will not permit the establishment of another Communist government in the Western Hemisphere.'

As the full cost of containment was being paid in Vietnam and as domestic criticism increased, the policy-makers sought to spread the costs of containment. The Nixon Doctrine announced in 1969 was the Truman Doctrine writ in an era of retrenchment. The doctrine stated that, while the USA would keep all its treaties, it would 'look to the nation directly threatened to assume the primary responsibility of providing the manpower for its defense'. The Nixon Doctrine was fighting the containment war by proxy. The USA would provide the aid and the hardware, and the client states, such as South Vietnam, would do the actual fighting. Regional client states were established to defend US interests in certain areas of the world. In the Middle East this role was performed by Iran under the Shah. The Carter Doctrine was announced in January 1980, in response to the Soviet invasion of Afghanistan. The doctrine indicated an overall increase in the defence budget and renewed emphasis on US commitments in the oil-rich states of the Middle East. Carter stated:

> An attempt by any outside force to gain control of the Persian Gulf region will be regarded as an assault on the vital interests of the United States. It will be repelled by use of any means necessary, including force.

Throughout the postwar years the USA has followed a path which has meant global involvement and global responsibilities.

THE USSR

The legacy of the past

When the Bolsheviks sought to capture state power in 1917 their ultimate prize was a huge empire which sprawled from Europe to Asia, and from the Arctic almost to India. It was an empire which abutted many countries and contained many different nationalities and ethnic groups. Russian expansionism in the nineteenth century was to leave its legacy on twentieth-century Soviet politics. The prize was not easily obtained and it was not until 1921 that the Bolsheviks secured victory against the Western powers and the White Russians.

The geo-politics of the country and the early experiences of the Soviet state influenced subsequent foreign policy. The feelings of vulnerability caused by the shared borders with so many, often hostile, neighbours were reinforced by the Western powers' attempt to strangle the socialist state at birth. In the civil war between 1917 and 1920 Britain, France and the USA supplied aid and troops to the Whites; in 1918 Bolshevik control was limited to a circle of 500-kilometer radius with its centre in Moscow. It is impossible to comprehend subsequent Soviet policy without noting its innate distrust of the world's capitalist countries. The warlike stance of these powers meant that Soviet policy was formulated from the position of a state encircled by aggressive nations.

The early years (1917–47)

Soviet foreign policy before the Second World War was one of caution in the face of perceived capitalist aggression. Care had to be exercised in defending the first socialist homeland and the emphasis was on pursuing peaceful co-existence and utilizing contradictions within the capitalist camp. The policy of peaceful co-existence was very largely determined by the backward state of the Russian economy. The country was shattered by the experience of the First World War (1914–18). It has been estimated that 16 million Russians died in the war against Germany and the ensuing civil war. A further 5 million had died of starvation and 2 million of typhus. The ravages of war imposed a conciliatory attitude towards the West. The term 'peaceful co-existence' was first coined by Lenin in 1917 as a realistic short-term response to the Soviet position. Thereafter, changes in foreign policy began to reflect the course of internal politics.

The body of doctrine which the Bolsheviks used to outline their policies was notably deficient in its coverage of the state. Marx had little to say on the politics of the state in capitalist society and even less on the role of the state in the transitional period. The Bolsheviks had to create their theories and their own state in the crucible of practice. In the initial period politics were determined by the obvious constraints. At the treaty of Brest-Litovsk in 1918 Russia signed an humiliating peace treaty with Germany in order to secure the life of the revolution. In the treaty Russia lost Finland, the Ukraine, Lithuania, Latvia and Estonia. Petrograd (which became Leningrad and is now St Petersburg) was so close to the border that the capital was shifted to Moscow. At home, the New Economic Policy was introduced to get the peasants to produce food to feed the starving urban workers. This policy, announced in 1921, brought the possibility of profits to peasants and returned small factories to private ownership. Private retail trade was also allowed and encouraged. As things began to improve, the economy got moving and most countries recognized the new state. By 1925 the possibility of alternative interpretations was raised. The tyranny of circumstances was beginning to lift.

The alternatives were raised in the leadership struggle between Trotsky and Stalin. Trotsky believed that socialism could not survive in one country, especially not in one as backward as Russia. The emphasis should be on encouraging world revolution, not only for its own sake but to ensure the continued existence of socialism in Russia. Only by encouraging socialist revolution in other lands at the time, combining their resources, knowledge and culture, could Soviet socialism prosper. Trotsky's belief in permanent revolution was Marxist internationalism writ large. For Stalin, in contrast, socialism in one country was a real possibility. Stalin believed that Europe would not rise in revolution. The function of Soviet policy was not, therefore, to spread revolution but to secure the existence of the Soviet Union. Trotsky offered further tribulations and pointed to Russian inadequacies while Stalin fused socialist rhetoric with gut nationalism. Trotsky's banishment in 1929 put the final seal on Stalin's victory and stamped the nature of subsequent policies.

The Comintern (Communist International) had been set up in 1919 in Moscow as a worldwide association of revolutionary Marxist parties. From its birth to 1923 it followed a policy of promoting revolution. Thereafter, with Stalin's victory over Trotsky and the failure of revolutionary movements in Europe, it became a creature of the Soviet Union Communist Party in general and Stalin in particular. From 1923 to 1928 the Comintern pursued a moderate line, a united-front policy in association with the view of the stabilization of capitalism. This policy changed as it came to be believed by Soviet theoreticians that capitalism was moribund. From 1928 to 1934 an ultra-left policy was adopted which committed communist parties to the direct conquest of power and the dismissal of all other political parties in capitalist countries. This policy led the German Communist Party to reject any idea of a socialist–communist coalition against the Nazis. Subsequent events showed this to be a disastrous mistake. It was only in 1935 that a popular-front policy was adopted. The emphasis was now on combining with other parties to form a platform of popular resistance against fascism. Even then, the Comintern had to change its policy overnight in August 1939, when Stalin concluded the Nazi–Soviet Pact.

Stalin's pact with Germany was an attempt to stave off an attack against the USSR. The German rearmament programme of the 1930s and German expansion into Austria and Czechoslovakia had brought only limited British and French response. When Britain and France signed the Munich Agreement with Germany and Italy in 1938 Germany was awarded part of Czechoslovakia as one of its 'legitimate claims.' Neither Czechoslovakia nor the USSR were present at Munich. For the Soviets then, Britain and France seemed content to deflect Nazi aggression towards the USSR. Soviet vulnerability was increased when the USSR and Britain failed to reach any agreement over policy in 1939. The Soviets therefore signed a pact with Germany in 1939. When the Germans invaded the Soviet Union in 1941, Britain and the USSR became allies in the fight against Hitler.

The cold war, phase one (1947–64)

The roots of the cold war lie in the hot war of 1941–5. From the Soviet perspective the USSR had borne the brunt of the war. Up until 1944 it alone had faced the German army in Europe and the Soviet losses of 20 million dead were the largest for any Allied country. The failure of the Allies to open a second front until 1944 was bitterly resented. Three things had been learned by communist policy-makers in the experience of war:

- do not trust the West
- the USSR, by virtue of its sacrifice, had earned a place amongst the world powers
- the new international order had to secure for the Soviet Union a good defensive buffer, access to sea routes and economic resources.

Conflict with the USA crystallized in Eastern Europe. Battle lines were drawn through the centre of Europe. For the Soviets the enunciation of the Truman Doctrine (1947), the implementation of the Marshall Plan (1947) and the creation of NATO (1950) were all acts aimed at them. They were all actions indicating a course of Western action leading to war and the destruction of the Soviet Union. The Soviet response was fourfold.

1. The setting up of Cominform (Communist Information Bureau) in 1947. At its inception the Soviet delegate Zhdanov outlined the two-camp analysis of the world order. The world, as pictured by Zhdanov, was divided into the socialist and capitalist camps. The division was marked by mutual hostility and conflict, with the USSR as a besieged camp hemmed in by non-communist forces. The two-camp analysis provided the basis for Soviet foreign policy until Stalin's death.
2. The Soviet Union tightened its grip on Eastern Europe. Comecon (Council for Mutual Economic Assistance) was set up in 1949 to pull the satellite states in Eastern Europe further into the Soviet orbit. Many of the satellites were stripped of movable goods and communist governments were set up which repressed any unrest. Repression became more pronounced in 1948 when Yugoslavia 'defected' from the Soviet camp.
3. The Warsaw Pact was established in 1955 after West Germany joined NATO. The Soviets were very wary of German rearmament and the Allied policy of bringing Germany into NATO was seen as an aggressive act which had to be countered.
4. Since Soviet attitudes to the Third World were refracted through the prism of the two-camp analysis there was very little appreciation of the independence of nationalist movements which did not toe the Stalinist line. You are either with us or against us, was the Soviet line; there could be no middle way.

Stalin's death in 1953 marked the beginning of a change in Soviet foreign policy. After a three-year transition period Khrushchev emerged as the leader and the new-look foreign policy was outlined at the famous Twentieth Congress of the Communist Party of the Soviet Union in 1956. Two principles were expressed:

- Emphasis was to be placed on peaceful co-existence with the West. The two-camp thesis, with its inevitable conclusion of war, was to be abandoned. The grisly reality of nuclear weaponry meant that the Soviet Union was unlikely to survive a full-scale atomic war. The rhetoric of war hardly fitted the Soviet Union given its tactical inferiority to the USA.
- Emphasis was to be placed on the periphery. The crude dichotomy of for and against, capitalist and communist, was abandoned. Socialism could be advanced by aiding nationalist movements even if they were not avowedly communist. The whole process of decolonization was opening up new opportunities for a change in the balance of world forces and the Soviet Union hoped to gain from these changes. It was at this time that aid

was given to, amongst others, Cuba, Egypt and India.

The thaw in Eastern Europe was initially signalled by the setting up of more equitable economic arrangements within Comecon and the first stirring of de-Stalinization. The relaxation of Soviet control gave nationalist communists in Poland the chance to gain control in 1956. Things began to get out of hand for the Soviets, however, in Hungary, where a popular uprising was brutally put down in 1956. After Hungary, communist party control tightened in Eastern Europe.

The new-look foreign policy suffered from a number of problems. There was a basic contradiction between the pursuit of co-existence with the West and the aiding of socialism in the periphery. The Soviets had to uphold the socialist cause

BOX F: SUMMARY OF USA–USSR RELATIONS: 1

In the post-Second World War era four separate periods of USA–USSR relations can be identified, periods which differ in terms of the nature of their relationship, the areas of tension and the methods used to pursue their geopolitical goals.

The first cold war, 1947–63/4

Very soon after the ending of the world war the two superpowers were locked into aggressive postures. Washington saw the USSR as a menace, a country which held control of Eastern Europe, did not allow either free elections or US trade and was bent on world domination. The Soviets posed a threat to US interests (often translated as world peace or democracy) and revolutionary movements in the periphery were seen through the prism of the cold war.

The immediate response was the Truman Doctrine, announced in 1947, which articulated the role of the USA as the world's 'policeman'. This rise of globalism involved a system of alliances and direct military involvement. The view from Moscow during the cold war was of the Soviet Union as a beleaguered country, hemmed in by a US-dominated system of alliances forming a noose around the Soviet state. Each saw the other's hand behind political conflicts throughout the world. Thus the State Department in Washington could 'explain' Vietnam and Cuba as 'communist revolution' inspired by Moscow, while Moscow portrayed the events of Hungary in 1956 as the result of 'CIA machinations'.

In the Third World, especially in South-East Asia and South America, any dictator or government losing support and facing popular revolt could blame it all on 'communism' to secure US intervention and aid. Through this process the US was sucked into maintaining unpopular regimes. US involvement in South Vietnam is a sad example. In Eastern Europe and Afghanistan the Soviet Union also maintained unpopular regimes. In this first period, then, the superpowers seemed locked into the rhetoric of irreconcilable conflict. The USA had a greater world role, enabling it not only to maintain collaborative elites in the area of the Monroe Doctrine but also establishing bases and troops in Asia and Europe. The USSR, in contrast, was limited in its direct interventions to Eastern Europe, an area where, despite the rhetoric, the USA made no threat to intervene. The Soviets had little direct impact outside this sphere of influence.

Round one of the superpower confrontation thus went to the USA with its superior military capability and greater global reach.

imminent in peripheral nationalist movements, but not to the extent of provoking massive retaliation from the USA. But to accommodate the West was to fail to uphold the socialist cause. The dilemma was further aggravated by the disparity between Soviet rhetoric and Soviet military capability. Under Khrushchev the Soviet Union was developing a global perspective but, despite Soviet claims to the contrary, the USSR had neither nuclear parity with the USA nor second-strike capability. In other words, the USSR could not strike back if first attacked by the USA. The USSR could attack the USA but she could not withstand the USA's retaliatory measures. Soviet claims to the contrary only made matters worse because they provided one of the more powerful reasons for the build-up of intercontinental ballistic missiles (ICBM) begun in the USA under the Kennedy administration.

The problems and contradictions became apparent in the Cuban missile crisis. Castro came to power in Cuba in 1959 on the wave of popular rebellion against the corrupt regime of Batista. Castro's agrarian reform provided cold comfort for US capital and when the Cubans signed a trade agreement with the Soviet Union in February 1960 the USA severed trading links. Cuba was now in the socialist camp. In 1961 the CIA-inspired Bay of Pigs invasion attempted to topple Castro and remove the socialist government. The attempt failed but it showed the lengths to which the USA was willing to go to overthrow the socialists. The Cuban government wanted protection from the US threat. For the Soviets, Cuba provided an invaluable base against the USA. On 14 October 1962, the US president was informed of the presence of nuclear missiles in Cuba. Kennedy imposed a naval quarantine to stop further shipments of arms and technology. In the phrase of the time, the world held its breath. After numerous diplomatic exchanges Khrushchev announced on 28 October that the weapon systems would be stripped and shipped back to the Soviet Union. The expansionist policy of the USSR had floundered and its global role was found wanting. Soviet vulnerability and inferiority to US missile power had been

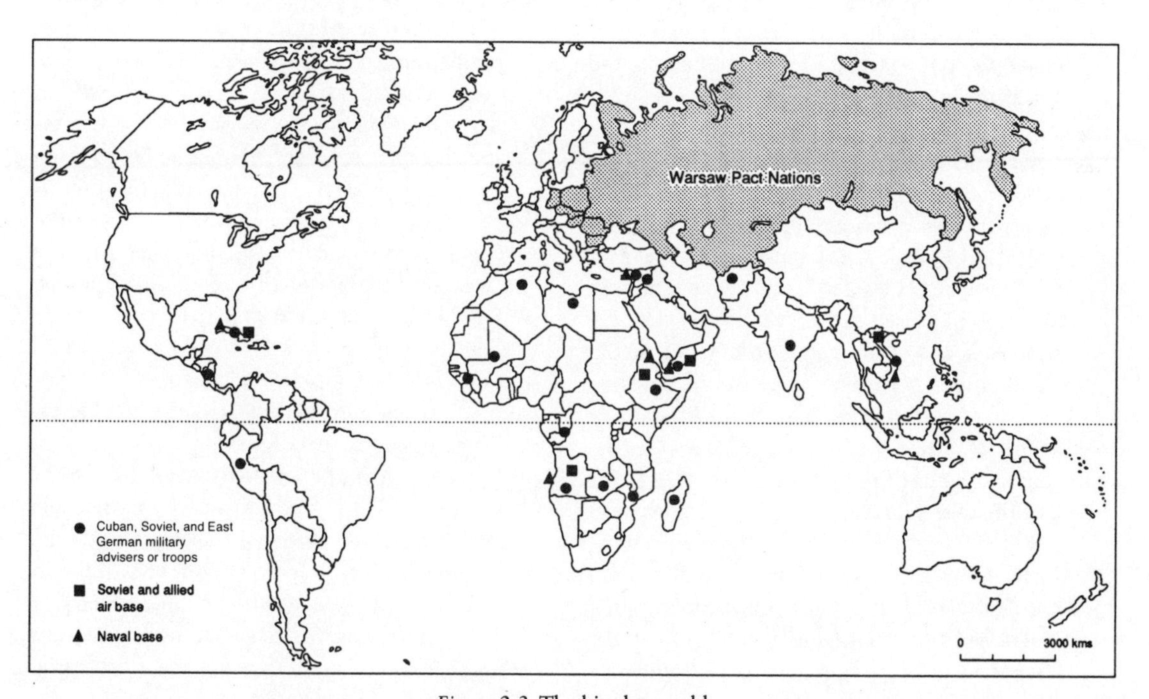

Figure 2.3 The bipolar world

demonstrated and further proof had been given to the Chinese argument that the Soviet Union was merely a 'paper tiger'; on the same day that Khrushchev was replaced as Soviet premier (11 October 1964) the Chinese exploded their first A-bomb.

The cold war, phase two (1964–85)

From 1964 until recently, Soviet foreign policy was based on achieving four goals:

1 the continuation of a global role for the USSR
2 the need to achieve peaceful co-existence with the USA and the West
3 maintaining hegemony in Eastern Europe
4 aiding socialism and socialist states in the periphery.

Let us look at these goals in more detail:

1 The global role of the USSR has been achieved by a massive build-up of arms, military power and strategic capability. Up until 1961 the USSR had only first-strike capability. In the 1960s a decision was made, perhaps in response to the Cuban crisis, to match the strategic power of the USA. By 1966 a second-strike capability had been obtained and by 1972 a rough parity in nuclear arms was achieved with the USA. In the 1970s the USSR had world-wide military mobility and a global strategic response (see Figure 2.3). The new military status brought some advantages. As the USA had used its power to keep friendly regimes in power throughout the world, so the USSR could now use its forces to aid socialist revolutions in the periphery. In Africa, for example, Soviet airlifts helped sustain the Angolan and Ethiopian revolutions against anti-socialist forces. And in Afghanistan Soviet troops were used to bolster a crumbling socialist regime. The military build-up also brought its problems. Anti-Soviet forces in the West used it to make the argument that the Soviets were bent on pursuing an aggressive stance against the West. The arms race began to spiral as NATO demands for arms strengthened the hand of the hawks in the Kremlin, which in turn aided the arguments of the hawks in the Pentagon. Each side reacted to the other; each side responded to the other's reaction. The global military role has also been expensive for the USSR. Greater military expenditure meant less investment in heavy industry and consumer durables. Thus it placed limits on economic growth and rises in consumer welfare.
2 There have been sound economic reasons for pursuing peaceful co-existence with the West. To obtain greater economic growth it was necessary to reduce military expenditure and get access to Western technology. Neither aim could be achieved while the Soviet Union was locked into an aggressive stance and a spiralling arms race with the West. The USA also had its own reasons for pursuing detente, especially the need to reduce military expenditure and the trade deficit.

   An increasingly important factor in the USSR's pursuit of co-existence with the West was the entry of China into world power calculations. From the time of the Cuban missile crisis the Chinese were convinced that the Soviet Union was revisionist and expansionist. The revisionist critique is seldom heard in post-Mao China but at the time it was a useful stick with which to beat the Soviet Union. Beneath the rhetoric the Chinese were fearful of a powerful state on their border; they felt that they had to distance themselves from the USSR if they were to be independent. Sino-Soviet relations deteriorated throughout the 1960s as the ideological disputes erupted in a series of border clashes. The Soviet fear of encirclement was reinforced when the USA, previously an implacable enemy of China, began a dialogue with the Chinese leaders. Nixon's visit to China in 1972 marked the end of the Chinese policy of isolationism towards the West and raised the Soviet fear of complete encirclement by hostile neighbours. Unable to overcome its disputes with China, and anxious to get Western technology, the Soviet Union became more eager to reach peaceful negotiations with the West.

3 Until recently a major goal of Soviet foreign policy was to maintain communist party hegemony in Eastern Europe. The aim was to foster the development of communist parties loyal to the Soviet Union. After the end of the war the nature of Soviet involvement in Eastern Europe was one of successive liberalizations followed by repressions, the cycle of events being synchronized with the changes in overall East–West relations. When the cold war intensified Soviet power hardened; during thaws the grip loosened. This was the overall trend, although it is important to bear in mind that the experiences of various countries have differed and that the freeze–thaw cycle has to be set against a trend of greater independence for Eastern European communist parties. There was an inexorable increase in the amount of leeway afforded to Eastern Europe. After 1964, as Soviet policy became characterized as a kind of flexible conservatism, greater freedom was given to Soviet satellites. Romania began to pursue a more adventurous foreign policy and Hungary began to inaugurate internal political reforms. These moves were halted by the 1968 Soviet invasion of Czechoslovakia. This took place because the Soviets were fearful that the Prague Spring of liberal reforms would sweep away communist party power in a country whose geo-political role was vital to the Soviet concept of a defensive buffer.

Czechoslovakia prompted the Soviet version of the Truman Doctrine. Three months after Soviet troops had replaced the Dubchek government with a more conservative communist regime, Brezhnev outlined the Soviet attitude in what later came to be termed the Brezhnev Doctrine. This asserted the right of the socialist community (i.e. the USSR) to intervene in the territory of any member of the community threatened by internal or external forces hostile to socialism. In effect, the Brezhnev Doctrine asserted the right of the USSR to maintain the communist party system of power in communist countries. In a bipolar world no country was to be allowed to break away from the socialist camp. The Chinese perceived the doctrine as a potentially dangerous policy for their independence and it strengthened the resolve of the Chinese leaders to distance themselves from the Soviet Union.

4 The final goal of the USSR was to aid liberation movements and fledgling socialist states in the periphery. The crude dichotomy of only non-communist and communist states was abandoned and the Soviets recognized a threefold division of the periphery into progressive socialist countries, nationalist independent states which included bourgeois elements, and capitalist regimes. Soviet policy in the periphery was aimed at helping the first two. Military assistance was tempered by the need to pursue peaceful co-existence with the West and by the lack of a global military capacity. In South-East Asia, although the Soviets did give assistance to the North Vietnamese, there was little likelihood of the USSR going to war with the USA over Vietnam. Even when the USA was involved in the invasion of Laos the Soviet response was muted. Soviet confidence, however, began to grow. The heavy investment in the military sector gave the USSR nuclear parity with the USA and global reach by 1972. Moreover, the debacle in Vietnam had caused the USA to temper its own interventionist tendencies. Soviet intervention grew in scope and depth against the background of a relative waning of US power and revolutionary upheaval in parts of the periphery. The growth was not uniform throughout the world. Little help was afforded to Allende's Chile and less for other South American states. The Soviet Union did not transgress into the clearly demarcated areas of the USA's sphere of influence but Soviet intervention grew in the grey, interstitial areas.

Intervention varied according to the circumstances. In Cuba the Soviet purchase of sugar kept the economy afloat. In Ethiopia the Soviet Union helped 'save' the revolution and aided the Ethiopian fight-back against the Somali invasion of the Ogaden. And in Angola Soviet airlifts helped defeat, for a time at least, the counter-revolutionaries and mercenary forces from South Africa. The commitment to intervention led to Soviet involvement in Afghanistan, which had implicitly been in the Soviet sphere of influence since 1956, when the country received $25 million worth of Soviet aid. Even before 1978 Afghanistan was the biggest single recipient of Soviet aid. In 1978 a small urban-based, predominantly middle-class, communist party overthrew the Daoud regime (see Halliday, 1980). The coup owed more to internal tensions in Afghan society than to Soviet expansionism. However, a socialist neighbour was warmly appreciated in Moscow and the Soviets stepped up their economic and military assistance to the new government. The new regime quickly got into difficulties. The counter-revolutionary forces were organized and received some aid from China, Iran and Pakistan. The internal situation worsened as factional fighting broke out. With its growing repression the new regime was whittling away its

BOX G: SUMMARY OF USA–USSR RELATIONS: 2

A period of detente, 1964–late 1970s.

The cold war was fuelled by national ideologies and in America by what President Eisenhower called the US military–industrial complex, a mixture of technological industries and military advisers who posed military hardware solutions to diplomatic issues. The USSR also had a military complex continually putting forward military interests.

Despite this pressure, the years of irreconcilable conflict began to soften in the early 1960s. The cold war was expensive, using up, as it did, valuable resources, and it was also dangerous. The Cuban missile crisis of 1962 showed how close the world was to a nuclear holocaust. Attitudes began to change and an Anglo–US–USSR treaty limiting testing of nuclear weapons was signed in 1963. There then followed a period of detente lasting up until the late 1970s. The period was still marked by direct intervention by the two superpowers: US military involvement in Vietnam and the Soviet invasion of Czechoslovakia in 1968 to put down the 'Prague Spring' which threatened to topple communist party dominance.

In terms of USA–USSR relations this second period marked the growth of both the rhetoric and the practice of peaceful co-existence. The tension between the two superpowers was reduced and the threat of direct confrontation seemed to be diminishing. In terms of respective geopolitical strategies the periods of detente saw a lessening in direct US foreign intervention as a foreign policy response to the post-Vietnam domestic political scene. Indirectly, however, trade patterns and aid disbursements helped to maintain the position of pro-US elites in the countries of Central and South America and Asia. In sub-Saharan Africa US involvement was minimal. There was no African equivalent of the Monroe Doctrine. In contrast, the period of detente saw an increase in the global capacity of the USSR, nuclear parity was achieved in the early 1970s, and the Soviets used this capacity to aid socialist movements in Africa.

Round 2 can be marked as a draw between fighters of near equal weight.

slender support. The Soviets believed that the government could not survive with Amin as the head of state. In fact, Soviet forces in Afghanistan were instrumental in replacing Amin with Babrak Karmal in December 1979. The immediate results are well known. Afghanistan was used by anti-Soviet forces in the West to promote increased military expenditures and the pursuit of a new cold war.

A NEW ORDER

A second cold war

By the end of the 1980s the old bipolar structure of the world order was coming to an end. It was not a smooth flow, an uninterrupted move from conflict to detente. There was a hiccup in the late 1970s and early 1980s as the relationship between the two superpowers began to deteriorate. After the debacle of Vietnam and the trauma of Watergate, US foreign policy was more muted. The latter half of the 1970s saw self-imposed limits on US intervention and a growing awareness that political change in the periphery was not necessarily the outcome of communist machinations. In the Ford administration Kissinger failed to get support for US intervention in Angola as the liberation movements came to power after the departure of the Portuguese. There were, however, very strong forces working against the curtailment of the USA's global role. Powerful economic interests had to be protected, the arms lobby hated the business vacuum of detente and the conservative–nationalist backlash in the wake of events in Iran and Afghanistan all propelled the USA towards a greater role.

The failure to get the US hostages out of Teheran marked the end of the Carter administration. The Reagan presidency, in its beginning, marked a change in US policy. Defence expenditure was increased. The rhetoric was a return to the 1950s, the USSR was the 'evil empire,' direct involvement was exercised in Grenada and Libya and desperately wanted by some in Nicaragua. Reagan wanted the USA to 'walk tall' and to 'carry the big military stick.' Behind the rhetoric was a nostalgia for the recent past when the USA was unrivalled in both economic and military power. What was especially frightening is what E. P. Thompson (1980) termed the 'logic of exterminism': the seemingly in-built tendency for the continual build-up of nuclear capability. There

BOX H: SUMMARY OF USA–USSR RELATIONS: 3

The second cold war, late 1970s–mid 1980s

In the mid to late 1970s there was a deterioration in USA–USSR relations, a second 'cold war'. The limitations to US power, as dramatically shown by the US hostages in Iran and the failure to halt Soviet involvement in Afghanistan, strengthened the hawkish element in US governments. The build-up of Soviet arms was seen as a direct threat. The increased military expenditure of the USA in turn gave added weight to those in the Kremlin arguing for more resources to be devoted to defence. With each superpower responding to the build-up of the other's arsenal there was a steady escalation in nuclear weaponry. The dialogue of detente was replaced by the rhetoric of cold war.

At the end of the third round the superpowers squared up to each other and things looked set for a new cold war.

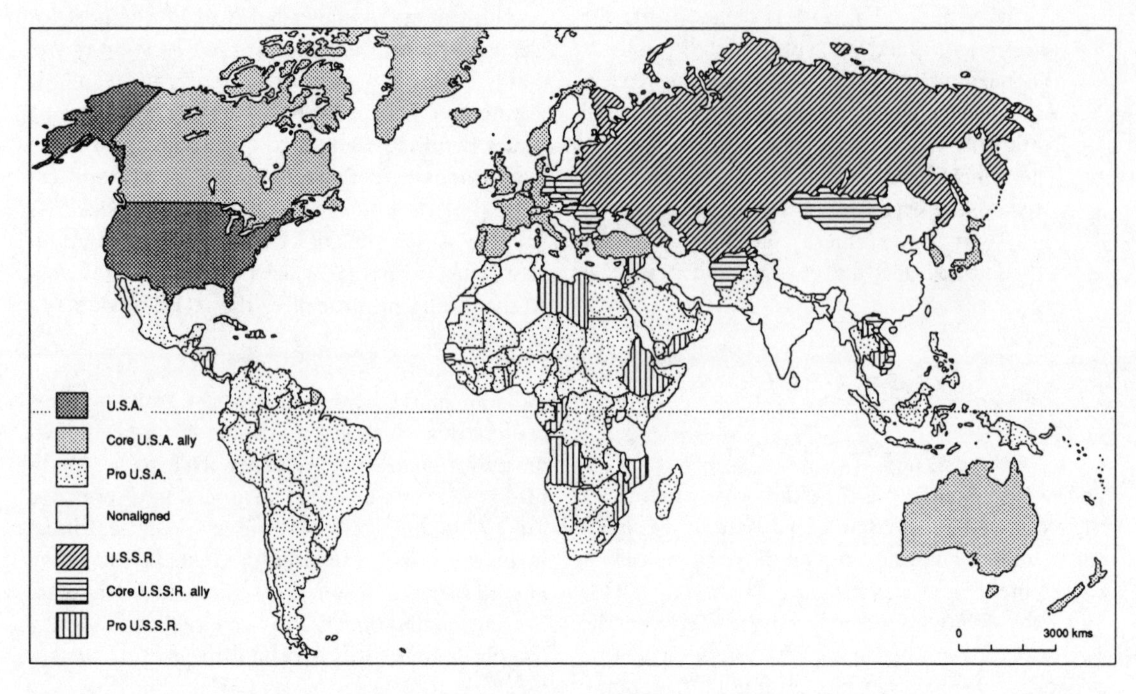

Figure 2.4 Superpower alliances, 1990

were powerful forces on either side of the iron curtain. Within the USA the economic impulses emanating from the military–industrial complex were transmitted through government to policies which have led to the continual build-up of arms stockpiles and recurring developments in nuclear missile systems. The military and economic forces were aided by nationalistic sentiments which justified the USA's possession and right to use nuclear weapons. There were equally powerful pressures within the Soviet Union, a society which had since 1917 been fearful of attack and been placed on an almost permanent war footing. The military and defence forces created to deal with the threat have their own momentum. The huge Soviet military–industrial complex was important enough to influence policy decisions, its inert presence shaping issues and guiding policies.

When the first edition of this book was written I entitled this section *Apocalypse now?* It included the following passage.

> The ultimate struggle between the two superpowers is embedded in the development of nuclear arsenals. Both countries now have the ability to destroy most of human life on the planet. The prospect of that apocalypse has evaded widespread discussion. Like our inability to face our own mortality, we find it difficult to contemplate the demise of the human race. It is a prospect so terrifying that we tend to ignore it.... The final point will be a nuclear exchange between the superpowers in which at least most of Europe will be fried to a radioactive mush, and at the most, 'A balance-sheet of the last two millennia would be drawn, in every field of endeavour and of culture, and a minus sign be placed before each total.' (Thompson, 1980, p. 29).

It was a period of pessimism, yet also one of action and hope. The peace movement grew, throughout the world. People took to the streets in New York and Moscow, and less than twenty miles from where I lived women attempted to barricade the

US military base of Greenham Common in Berkshire. When the history of the momentous changes come to be written, the importance of such bottom-up pressures should not be forgotten.

Figure 2.5 The world centred on Moscow

Figure 2.6 The world centred on Washington

It seemed darkest before the dawn. By the late 1980s the world was divided up by the superpowers (see Figure 2.4) and there were two opposing world views, one from Moscow and another from Washington (see Figures 2.5 and 2.6).

All change

When Gorbachev became General Secretary of the Communist Party of the Soviet Union in 1985 most people thought he was just another grey figure in a long line of grey figures. How wrong can you be! Gorbachev sought reform. At home he wanted openness (*glasnost*) and a restructuring (*perestroika*) of Soviet society. On the world diplomatic scene he became the leading figure seeking a new dialogue between the superpowers. In November 1987 the Intermediate Nuclear Forces Treaty was signed. This treaty involved the disarming of missiles in Europe, East and West. The treaty also marked a significant change in US attitudes; the Soviet Union was no longer the evil empire. By 1992 it was no longer even an empire.

The decline of the bipolar structure raises new opportunities. It also raises problems as old superpowers seek new roles. The USA no longer has the USSR to give coherence to a simplistic world view and the Russians have lost their empire. Apart for so long, the USA and Russia may well share the same withdrawal symptoms, like two old actors, former stars, unwillingly leaving the stage they have dominated for so many years, not quite sure what to say or do. Their old lines, their former roles, are no longer appropriate.

The new world order

In 1990–1 a new term was heard in international diplomatic circles: *the new world order*. Like all 'new' terms it was not all that new, as politicians have been predicting a new world order for at least a couple of centuries. The US political geographer Isaiah Bowman wrote a book entitled *The New World: Problems in Political Geography*. The book was first published in 1922. And like most fashion-

BOX I: THE EVIL OTHER

Empires tend to come in pairs; they struggle and vie with each other for domination. Athens and Sparta, Rome and Carthage, the USA and the USSR; the bipolarities continue through the centuries. An essential element in their relationship is their denigration of each other. The competitor becomes the 'evil other', the source of disorder and unrest, a country populated by demons and devils. In the rhetoric of the cold war the USA saw the USSR as a menace to peace and world harmony, hell-bent on world domination. The Soviets, in contrast, saw an enemy empire which had military bases all over the world, which had used the atomic bomb on innocent civilians, which wanted to destroy their society. Opinions were polarized; it was a case of forces of good against the power of darkness. The ideologies fed off each other, they needed each other to provide an enemy, an easily identifiable source of trouble. The USA could blame the USSR for social unrest around the world, the USSR could see the hand of the Americans whenever the population in their satellite states of Eastern Europe wanted more independence. There was a symmetry. The CIA could see the KGB at work, the KGB was sure of CIA involvement. Military build-up in the USA led to a military build-up in the USSR, which led the Pentagon to ask for more money, which in turn led the generals of the Red Army to demand more military hardware.

In 1987 the American writer Norman Mailer visited Moscow. He was reported as saying:

> I've come here to tell you that you must continue with *glasnost'* for at least two years. It will take that long for the US media to figure out what's happening here and to design a way to communicate it to the US public. Of course the Americans have a big problem here. The moment we cease to think about the USSR as the evil empire, we have to think that the enemy is elsewhere, possibly even in ourselves.
>
> (Sheff, 1987)

able terms it was used to mean a rich variety of things. Its usage, however, did mark the recognition of a huge change in the world order, and reflected three trends.

1 The decline of the USSR as a superpower. Racked by economic difficulties, which made it no longer able to sustain a global military presence, and riven by internal dissensions, which made political consensus impossible, the USSR was no longer a superpower able to compete with the USA. As the loosening of political repression opened up a host of centrifugal forces and the ethnic mosaic finally began to unravel, the USSR collapsed. The old bipolar structure which had dominated the post 1945 world was, by 1990, a thing of the past. Now it is only a subject for historians.

2 The decline of the USSR means that the USA is the undisputed military power in the world. Its global reach is unmatched by any other single country. After 40 years of sparring with an opponent, the USA has emerged the winner. The 1991 Allied victory against Iraq in Kuwait showed the awesome might of US firepower and missile technology. However, as the 1991 Kuwait campaign also showed, for US power to be successful some form of legitimacy is essential. The USA cannot wield its power independently of global opinions. Ensuring international support is vital if the USA is to be a credible world leader. This involves a set of alliances, some of which may involve unusual combinations. The Kuwait campaign was unusual, bringing together the USA, France and Britain as well as Egypt,

Saudi Arabia and Syria, but in the new world order new systems of alliances, combinations of a more subtle, pragmatic and flexible kind may be more the norm than the exception.

3 While the USA retains its dominant military position it no longer has undisputed economic power. The rise of Japan, the development of Germany, as well as the competition from the EC, have all made inroads into the domestic and foreign markets of the USA. The undisputed military power does not have undisputed economic power. The Germans and Europeans may have less military might but their fiscal power is very strong. In the case of the Kuwait campaign, the USA did most, but not all, of the fighting but sought to spread the economic burden to rich, non-combative Japan and Germany. The new world may be more politically stable, but it is ever more economically competitive.

By 1992 the world was a very different place from what it was in 1950. The USA could no longer see revolutionary upheavals around the world as a co-ordinated Kremlin plot. The world, at least as perceived by the more intelligent people in Washington, could no longer be seen in the simple terms used during the cold war. Upheavals in Latin America, coups in Africa and rioting in Asia could not be easily explained or understood as a Soviet-inspired plot. A more sophisticated analysis is required from Washington, a greater awareness of the internal dynamics of social change and a greater sensitivity towards different types of societies around the world. The new world is more complex than the old world.

GUIDE TO FURTHER READING

Introductions to the foreign policies and geopolitics of the two superpowers include:

Agnew, J. A. (1983) 'An excess of national exceptionalism: towards a new political geography of American foreign policy'. *Political Geography Quarterly* 2, 151–66.

Ambrose, S. (1985) *Rise to Globalism: American Foreign Policy Since 1938.* Penguin, Harmondsworth.

Chomsky, N., Steele, J. and Gittings, J. (1982) *Superpowers in Collision.* Penguin, Harmondsworth.

Halliday, F. (1990) *Cold War, Third World. An Essay on Soviet–American relations.* Hutchinson, London.

Kolko, G. (1988) *Confronting the Third World: US Foreign Policy 1945–80.* Pantheon, New York.

LaFeber, W. (1985) *America, Russia and the Cold War.* Knopf, New York.

Pugh, M. and Williams, P. (eds) (1990) *Superpower Politics: Change in The United States and The Soviet Union.* Manchester University Press, Manchester.

Sloan, G.R. (1988) *Geopolitics in United States Strategic Policy 1890–1987.* Wheatsheaf, Brighton.

Steele, J. (1983) *World Power: Soviet foreign policy under Brezhnev and Andropov.* Michael Joseph, London

Van Alstyne, R.W. (1960) *The Rising American Empire.* Oxford University Press, Oxford.

The geography of their foreign policy is best considered in maps. Have a look at the following atlases:

Barraclough, G. (ed) (1978) *The Times Atlas of World History.* Times Books, London.

Barnaby, F. (ed) (1988) *The Gaia Peace Atlas.* Pan, London.

Bunge, B. (1988) *Nuclear War Atlas.* Basil Blackwell, Oxford.

Chaliand, G. and Rageau, J.P. (1985) *Strategic Atlas of World Geopolitics.* Penguin, Harmondsworth.

Freedman, L. (1985) *Atlas of Global Strategy.* Macmillan, London.

Kidron, M. and Smith, D. (1983) *The New State of The World Atlas.* Pan, London.

Kidron, M. and Smith, D. (1983) *The War Atlas: Armed Conflict – Armed Peace.* Pan, London.

On the more recent changes, for the USSR consider:

Crouch, M. (1989) *Revolution and Evolution: Gorbachev and Soviet Politics.* Philip Allan, Oxford.

Medvedev, Z. A. (1988) *Gorbachev.* Basil Blackwell, Oxford.

Nove, A. (1989) *Stalinisation and After: The Road to Gorbachev.* Unwin Hyman, London.

Rowen, H.S. and Wolf Jur, C. (eds) (1988) *The Future of The Soviet Empire.* Macmillan. London.

White, S. (1990) 'Democratization in the USSR', *Soviet Studies*, 42, 3–25.

And for the USA see:

Calleo, D. P. (1987) *Beyond American Hegemony: The Future of The Western Alliance.* Basic Books, New York.

Friedberg, A.L. (1989) 'The strategic implications of relative economic decline', *Political Science Quarterly*, 104, 401–31.

Relevant journals

Dissent
Foreign Affairs
Foreign Policy
International Affairs
International Studies Quarterly
Journal of Strategic Studies
Monthly Review
Soviet Studies
Survey
World Policy Journal
World Politics

Data sources on military strength include:
Adelphi Papers
Military Balance
Stockholm Peace Research Institute (annual surveys)
Survival

Other works cited in this chapter

Bowman, I. (1922) *The New World. Problems in Political Geography*. Harrap, London.

Halliday, F. (1980) 'The war and revolution in Afghanistan', *New Left Review*, 119, 20–41.

Magdoff, H. (1969) *The Age of Imperialism.* Monthly Review Press, New York.

Sheff, D. (1987) 'Back in the USSR', *Observer Magazine*, 3 May.

Thompson, E. P. (1980) 'Notes on exterminism, the last stage of civilization', *New Left Review* 121, 3–31.

Thompson, E. P. (1990) 'When the war is over', *New Statesman and Society*, 26 January, 19–31.

Turner, F. J. (1963) in H. P. Simonson (ed.) *The Significance of the Frontier in American History.* Ungar Publishing: New York.

Williams, W.A. (1961) *The Contours of American History*. The Viewpoints. New York.

3

THE MULTIPOLAR WORLD

> The American Empire which, like most empires, was as much an empire in men's minds as an empire over palm and pine, reposed in the Old World upon two pillars. The one pillar was the proposition that Soviet Russia and its East European allies were bent upon the invasion and conquest of western Europe. The twin pillar was the proposition that the invasion had been averted and still continued to be averted, by the determination of the United States to react to it by committing nuclear suicide.
>
> Considering that both propositions are contrary to reason and observation, it is remarkable that the two columns remained in place for close upon 40 years before collapsing. But that is something apt to happen where there is wide-spread political vested interest in suspending incredulity.
>
> (Enoch Powell, 1988)

The political geography of the world order has, until very recently, been dominated by the actions of the two superpowers. The USA and USSR have dominated the world – indeed, to all intents and purposes they *were* the world order. They provided the context in which other nations adopted foreign policies and pursued economic objectives. In recent years, however, this bipolar structure has begun to change. The global dichotomy with its attendant allies and satellites has begun to break up. At times the change has been almost imperceptible, at other times events have moved very quickly. The general picture is of the absolute decline of the USSR, the relative decline of the USA and the emergence of new centres of power. We will examine these new centres in this chapter. First, however, let us consider the retreat from empire of the two old superpowers.

RETREAT FROM EMPIRE

On Monday night, 16 October 1989, a group of East Germans demonstrated against their government. They congregated in the main square of Leipzig and, to show their disapproval of their government, they chanted 'Gorby, Gorby'. Forty, thirty or even five years previously such an event would have been impossible. Dissident East Germans chanting the name of the Soviet Premier as an act of revolt against their own government! Hadn't the Russians sent the tanks into Hungary and Czechoslovakia for less?

To publish your writings is to commit your thoughts to paper. The danger is that you can be overtaken by totally unexpected events. In the first edition of this book, I wrote:

> What remains clear, however, is that the Soviet Union will not allow the development of forces leading to the ousting of the communist parties. The Brezhnev Doctrine and Soviet troops underwrite communist party power in Eastern Europe.
>
> (Short, 1982)

At the time of writing, and even up until 1989, every respected analyst would have agreed with that statement. However, the year 1989 marked a sudden change. Communist party power in Eastern Europe was swept away in a wave of popular protest (see Table 3.1). Behind these events lay two structural features:

1 *The retreat from empire by the Soviet Union.* The Soviet leader refused to send in troops when requested by the East German leader Erich Honecker in 1989. If anything, Gorbachev encouraged the popular uprising. The lack of Soviet intervention gave the green light to the open display of public dissent. The decline of bipolarity in the world scene meant collaborative elites in Eastern Europe could no longer use the tried and tested formula of the 'evil other' to explain and dismiss internal dissension. The old leaders could no longer appeal to Moscow or use the excuse of CIA involvement.
2 *The lack of genuine support for the communist governments.* Allowed the opportunity to express their feelings, the people showed an overwhelming desire to get rid of the old leaders and scrap the communist system.

The structure of traditional communist party power in Eastern Europe has been demolished. What will come in its place? We may see some countries opting for a market system while some may choose a more mixed public–private economy. What remains clear is that there is no support for a return to centralized undemocratic socialism.

Table 3.1 Events in Eastern Europe in 1989

Date	*Event*
4 June	Solidarity wins free election in Poland
19 September	Hungarians set date for free elections
9 November	Berlin Wall falls
10 November	Bulgaria's leader is overthrown
24 November	Communist party leadership of Czechoslovakia quits in the face of massive popular protest
5 December	Former East German leader is put under house arrest
22 December	Nicolae Ceauşescu toppled in Romania

The first half of the twentieth century has seen the tremendous development of the USA from world power to superpower. The story of the second half has been one of relative decline. (Chapter 4 considers some of the major reasons.) The USA is still the undisputed military superpower and the single biggest economy in the world. However, as the twentieth century was drawing to a close its economic power was threatened by a new European power centre and the emergence of Japan as an economic superpower.

In 1941 the publisher Henry Luce wrote, in a magazine article, of 'the American century'. The twentieth century was to see the full flowering of American power and global influence. In the immediate post-war world this was undoubtedly true. One crucial element which gave coherence, stability and purpose to the USA was the existence of the Soviet empire. The decline of the Soviet empire and the end of the communist threat puts the USA in a difficult transitional role. No longer is there the evil other, the countervailing world force. It becomes more difficult to prop up undemocratic regimes, when the communist threat has disappeared. The decline of the Soviet empire also heralds the decline of the old American empire.

In the new world order of the 1990s the bipolar structure of East versus West no longer has a place. Indeed, there is a growth of new power centres in the world. The bipolar world is being replaced by a multipolar world. In the rest of this chapter let us consider three important alternative centres of power (see Figure 3.1) and their emergence, especially since 1945.

CHINA

China is the most populous nation on earth. Its population of 1000 million constitutes almost one-quarter of the world total. If its resources were utilized as fully as those of the USA, then China the world power would become China the superpower.

China's evolving relationship with the USA and USSR has been complex. From the 1949 revolu-

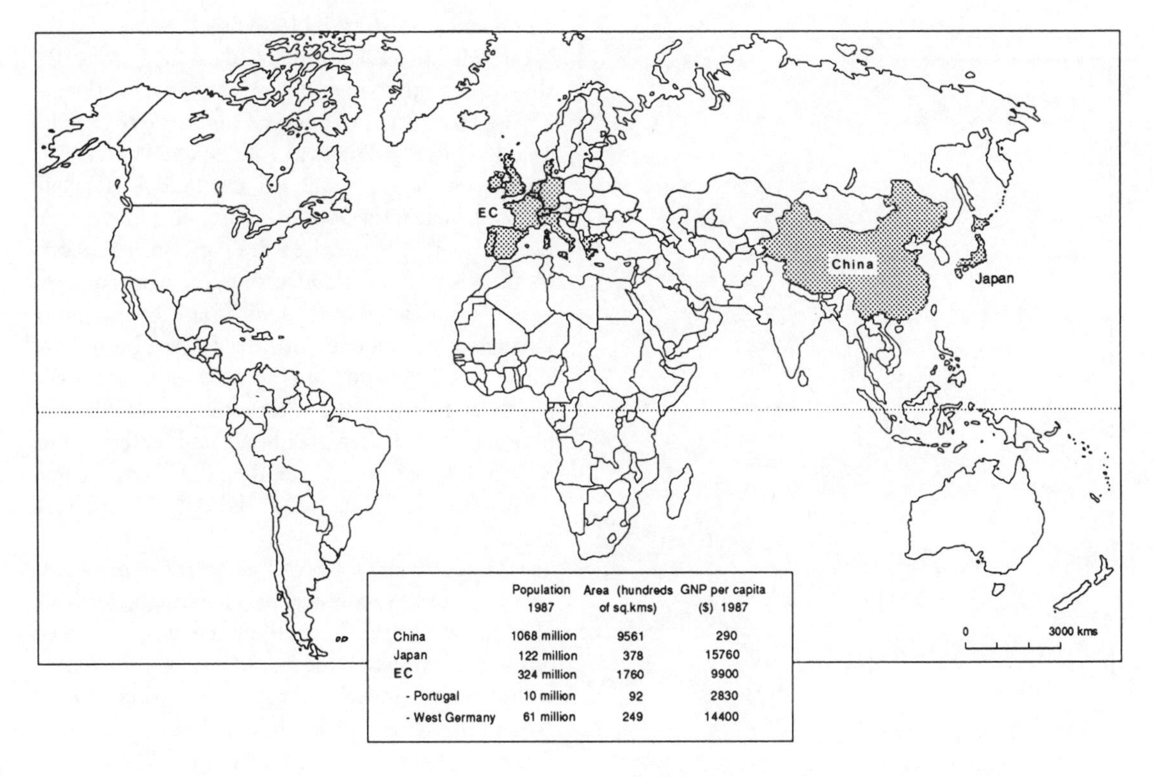

	Population 1987	Area (hundreds of sq.kms)	GNP per capita ($) 1987
China	1068 million	9561	290
Japan	122 million	378	15760
EC	324 million	1760	9900
- Portugal	10 million	92	2830
- West Germany	61 million	249	14400

Figure 3.1 A multipolar world

tion throughout most of the 1950s the USSR and China were allies against the cold war threat of the USA. The alliance was unequal. The Soviet Union was the more economically advanced country and laid claims to lead the world communist movement. China was to be placed in the role of junior partner. With Soviet aid and technical assistance, Chinese development began to follow the Stalinist path of centralized planning and a concentration of investment in heavy industry. Throughout the decade such policies resulted in a dislocation of the economy as the heavy industrial sector expanded more than the agricultural sector. The economic policies, according to Mao, were leading to growing disparities between rural and urban areas, manual and mental workers, and the party and the masses. Mao took the leftist view that the nature of economic growth was as important as the size of growth and succeeded in launching China into the Great Leap Forward in 1958. The leap attempted both to increase growth and foster Chinese independence. The disparities were to be reduced and eventually eradicated. The independence line in domestic policy was matched in military affairs. In 1958 the Chinese wanted a share in the command of Soviet bases inside China. The arguments continued until the USSR reduced economic support and withheld technical assistance. In 1960 the Sino-Soviet split was officially recognized when the USSR withdrew all forms of aid. Beneath the issue of the bases lay deeper differences in the interpretation of their alliance, border rivalry and old prejudices.

Throughout the 1960s the split widened. For the Chinese the Brezhnev Doctrine and the events in Czechoslovakia were sharp reminders of the power and determination of the Soviet leaders to maintain their hegemony. After the end of the Cultural Revolution in 1969, the Chinese leadership began to look towards the USA, Western

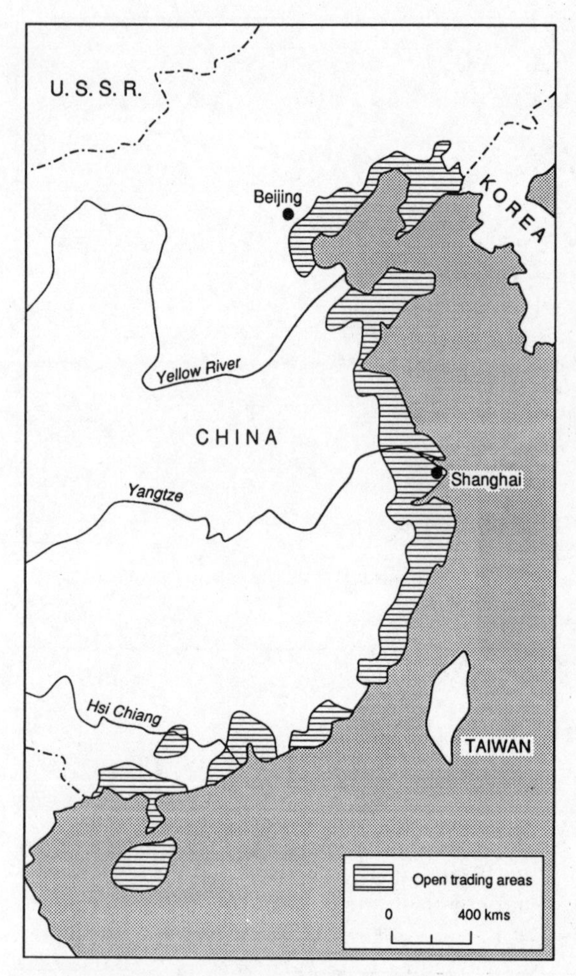

Figure 3.2 Open trading areas in China

Europe and Japan because they had the technology and capital necessary for economic growth. Links with the USA were encouraged in order to provide some measure of defence against the Brezhnev Doctrine being applied to China. The fostering of Sino-American ties would also stop China from being isolated in the climate of growing detente between the two superpowers.

From 1960 onwards the Soviet Union became wary of China. The old Soviet fear of encirclement was one reason for pursuing detente with the USA in the 1970s with such vigour; the Soviet nightmare was of facing two enemies, one in the East and one in the West. By lessening tension with the West the Soviet Union hoped to neutralize the threat of fighting on two fronts at the same time.

For the USA, links with China proved a valuable bargaining chip with the Soviet Union. The pursuit of detente widened the Sino-Soviet split and weakened the world communist movement. Detente with China raised the old hopes of access to Chinese markets and resources, while detente with the Soviet Union made the Chinese more amenable to dialogue with the USA. Throughout the 1970s the situation evolved into improved relations between the USA and the USSR and between the USA and China, and deteriorating relations between the USSR and China, though more recently China and the USSR sought a rapprochement.

China's leaders have sought to expand the economy of the country by encouraging private markets. Since 1979 economic reforms have been introduced which replaced Mao's vision with a willingness to trade with the outside world. Special economic zones were founded in Shenzhou, Zhuhai, Shanton and Xiamen in 1984 and 14 ports were declared open cities. In 1988, 140 additional ports were included as open economic zones (see Figure 3.2).

The leadership of China sought to introduce capitalism but limit democracy. This has proved a difficult task. The economic reforms meant rising inequalities and open corruption amongst the party bureaucracy. This led to popular discontent especially amongst the university students whose standard of living did not improve. In April 1989 students staged a sit-in in Beijing. It was broken up by the police Many people responded to this initial demonstration and a week later more than 150 000 people demonstrated in Tiananmen Square. The protests continued until the night of 3 June when the People's Liberation Army fired on the people. If East Germany in 1989 was a success for people's power then Tiananmen Square was its failure.

JAPAN

In the post-war period Japan has emerged as one of

the strongest economies in the world. The pace and scale of economic development has been impressive. Since 1960 Japan's growth rate has been almost double that of other major capitalist economies (see Table 3.2).

Table 3.2 Economic growth rates

	Annual average growth of GNP per capita 1960–77	1965–87
Japan	7.7	4.2
France	4.2	2.7
West Germany	3.3	2.5
UK	2.5	1.7
USA	2.4	1.5

Source: World Development Report, 1989, Table 1.

Japan began the post-war period under the control of the USA. Independence was only attained in 1952. From then until 1964 Japan was integrated into the US scheme of things. The USA provided defence, trade and capital. It provided two-thirds of Japan's imports and took three-quarters of Japan's exports. US multinationals became the major foreign capital investor. The Japanese economy received initial stimulation from the USA through the use of special procurements at the time of the Korean War. Thereafter, the USA actively encouraged the expansion of the Japanese economy and up until 1964 Japan was the second highest recipient of World Bank loans, which by that time had totalled $1500 million. For the USA a strong Japan provided an anti-communist bulwark against Soviet and Chinese influence in East Asia. Japan was on the perimeter of the 'free' world's defences, it was a bastion that had to be defended.

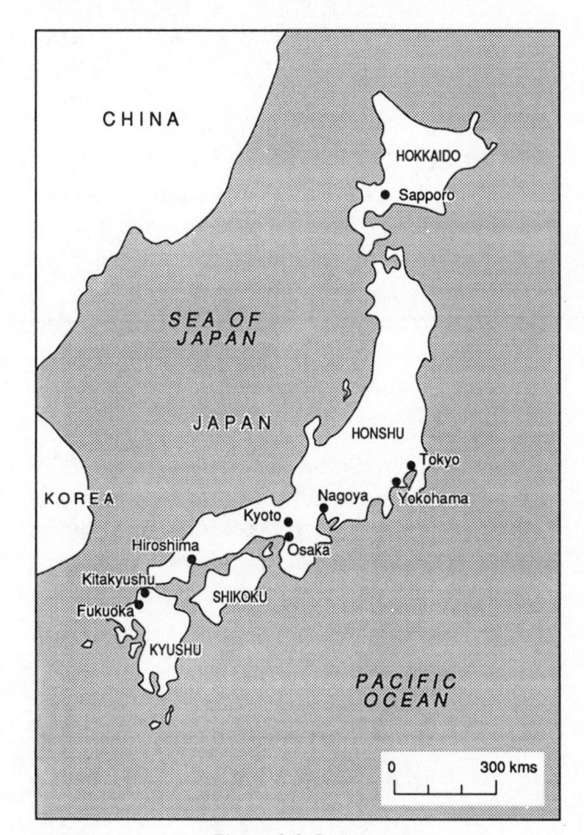

Figure 3.3 Japan

The post-war growth in the Japanese economy was not entirely due to US influence. Internal factors played an enormous role. Successive Japanese governments have been committed to economic growth and the high rate of capital accumulation has been achieved through the interlocking relations between *Zaikai* (big business and finance) and a state committed to a regressive tax system and very low levels of public expenditure (in the early days at least) on social services and welfare programmes. Economic growth has also been based on a strong domestic market. The main stimulus to production has been given by the home market, and even as late as 1980 over 50 per cent of cars produced in Japan were sold in Japan. As the Japanese consumer has got richer, the increase in effective demand has stimulated the economy and as economic growth has led to increases in wages, the benign spiral of economic growth has continued upwards. Several commentators have also drawn attention to the cultural matrix of Japanese society which has facilitated economic development (Kahn, 1971; Vogel, 1979). The cultural base of Japan has provided a rich humus for continued economic growth.

Japan's strong position in the world economy and her accelerated growth since 1964 has placed the country in a web of political and economic relations with other countries. As Japan's economy has grown, so has the need to import raw materials and export manufactured goods. Japan is very dependent on other countries for basic raw materials; for example, over 90 per cent of her oil comes from abroad, almost 75 per cent from the Middle East. This huge dependence on overseas supplies has necessitated a number of political strategies. Japan has cut her diplomatic cloth to suit her economic needs. This has meant cultivating a careful policy towards oil-producing countries in the Middle East and a continuing dialogue with the USSR concerning mutual exploitation of Siberia's natural resources. At times (e.g. during the early days of the Iranian crisis in 1979, when Japan continued to import oil from Iran after the US embassy hostages had been taken and the USA had called for a total trade embargo) the policies have drawn Japan in different directions from the USA.

The growth of the Japanese economy has been based on the export of manufactured goods. From the 1960s Japanese firms began to sell more of their goods both in the markets of the periphery and in Western Europe and North America. In the latter two areas the more efficient Japanese companies have undercut the less efficient US and European companies. Japanese market penetration into the USA and Europe has involved the closure of factories and the folding of companies. Such economic processes and their social implications have brought the political allies into economic conflict. The arguments have involved demands for protection against Japanese goods and demands for opening up the Japanese market to foreign firms. More recently, there have been strong demands in the USA for direct Japanese investment – to take the most obvious case, the construction of Japanese cars in plants located in the USA. During periods of world-wide economic growth the conflicts between Japan and the other major capitalist economies are not apparent. During depression and recession, however, they become acute as Japanese growth is purchased at the expense of other capitalist countries (see Table 3.2).

Japan is now one of the most important single national economies in the world. Her companies buy enormous quantities of raw materials from around the world and sell their goods across the globe. The strength of her industry has led to huge financial surpluses and the consequent growth of the Japanese financial system. In 1987 the Tokyo Stock Exchange overtook Wall Street as the single biggest financial centre in the world. The ten biggest banks in the world are now all Japanese. Japan is also one of the biggest overseas capital investors. All of these trends are likely to continue and Japan's role in the world economy will continue to grow. This growth will take a number of forms:

- the direct control of raw material supplies
- the search for cheap labour sites for routine manufacturing processes (e.g. Thailand, South Korea)
- the siting of highly polluting industries outside Japan and 'tariff-hopping' investments in Western Europe and North America. Japan's economic stability may therefore involve more explicit political involvement in international affairs. In South-East Asia radical commentators and many political parties see Japan as the imperialist power; in buying cheap raw materials, selling manufactured goods and using cheap local labour in routine manufacturing factories Japan is indeed taking over the classic economic–imperialistic role previously filled by European countries and the USA.

Japan's growing economic position in the world order provides the basis and the necessity for a greater political, and perhaps military, role in world affairs. Its economic success provides a potent source of tension with its allies in North America and Europe.

WESTERN EUROPE

Europe ended the Second World War in bad shape. Germany had been devastated and even the European victors were close to bankruptcy. With

Europe fracturing along the fault of the iron curtain, the USA began to think of resuscitating the economies of Western Europe. A strong, integrated Western Europe was seen as a vital basis of defence against the Red Army. It would also buy US goods and open up the world trading system and thus stave off the possibility of recession. There were also strategic and political reasons. Stronger European powers would be able to hold on to those colonies and territories which contained valuable raw materials and which were in danger of being lost to the communist camp. And economic growth in Europe would counter the claims of the radical left, especially in France and Italy where the communist parties were particularly strong. The US aims of a stronger, integrated Western Europe were manifested in the establishment of the Marshall Plan and the Organization for European Economic Co-operation (OEEC) in 1947 and of NATO in 1949. NATO was the defence arrangement, the Marshall Plan was designed to prime the economic pump and the OEEC was the framework for economic revival established to distribute the Marshall Aid monies. The OEEC reduced import controls and subsidies for exports and allowed the free movement of goods. It was the US Open Door policy applied to Europe. Later, the OEEC was replaced by the OECD (Organization for Economic Co-operation and Development) in 1961 and was expanded to include Canada, Japan and the USA.

The OEEC, OECD and NATO were organizations which provided a weak glue which loosely bound Western European countries. The leading light and guiding hand in the arrangements was the USA. Another strand in the move to integration was the creation of European organizations by the Europeans themselves. The forces of integration were many, ranging from the idealists eager to avoid the repetition of war to the businessmen eager to establish larger markets. The earliest creation was the Benelux Agreement, initially signed in 1944 but coming into force in 1948. This agreement reduced tariffs between Belgium, Luxembourg and the Netherlands. Next came the Council of Europe, set up in 1949 as a broad-based organization which aimed to 'achieve greater unity and facilitate economic and social

BOX J: NO LONGER A YEN FOR JAPAN

Former allies can very quickly become objects of fear and loathing. In the 1960s, when the USA was all-powerful, Japan was seen by most commentators as a valued ally. There were still some who held lingering resentment about Pearl Harbour and Japanese conduct during the Second World War, but the official line was that Japan was a good friend whose economic growth rates and industry were to be admired.

By the 1980s, however, attitudes had changed. Japan was now the world's economic superpower whose electronic goods and cars were sold throughout the world and whose money was invested across the globe. The rise of Japan was paralleled by the relative economic decline of the USA. For many Americans the two were connected. Japan was both admired yet feared, as the political ally became the economic enemy. Even when Japanese money was invested there were fears that Japanese companies were disrupting normal American work practices and that Japanese money was loosening American control of American resources. The debates were curiously similar to the European fear of American investment in Europe in the immediate post-war era. For a sample of some of the more xenophobic literature have a look at:

Burstein, D. (1990) *Yen. Japan's New Financial Empire and Its Threat to America.* Fawcett Columbine, New York.

Wolf, M. (1983) *The Japanese Conspiracy.* Empire, New York.

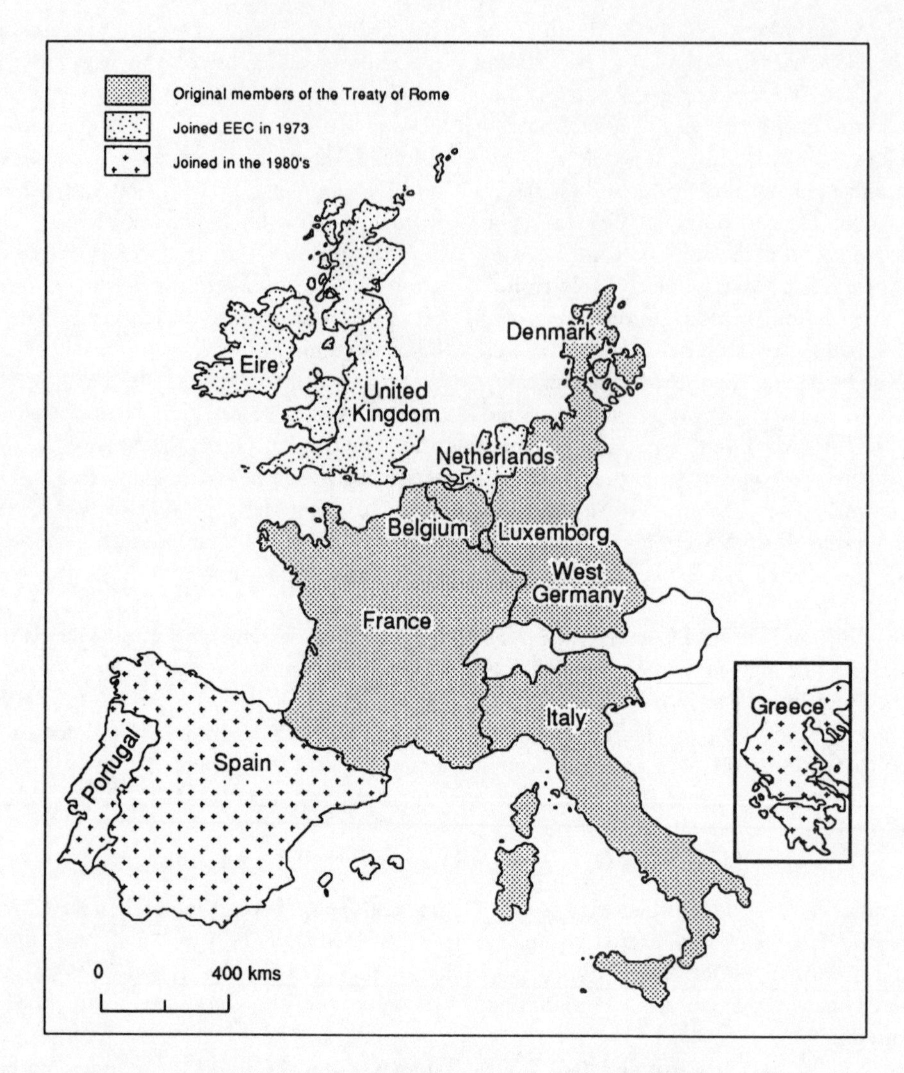

Figure 3.4 The European Economic Community, 1992

progress'. The woolly programme and vague policies meant that it was little more than a talking shop for bureaucrats and diplomats. The most significant development in the early post-war years came with the setting up of the European Coal and Steel Community (ECSC) in 1952. The ECSC included Belgium, West Germany, Italy, Luxembourg, the Netherlands and France. It was to lay the foundation of the European Economic Community (EEC). The ECSC was initially designed to control German steel and coal production. During the Korean War there were heavy demands for steel, but the only spare capacity in Western Europe was in West Germany. The French, and others, were naturally worried by the prospect of a revival of German steel production, a sector traditionally associated with armaments. The Schuman Plan, drawn up in 1950, offered a way of promoting German steel production but giving control to France and other European nations. The ECSC grew from this suspicious context to an organization for planning and

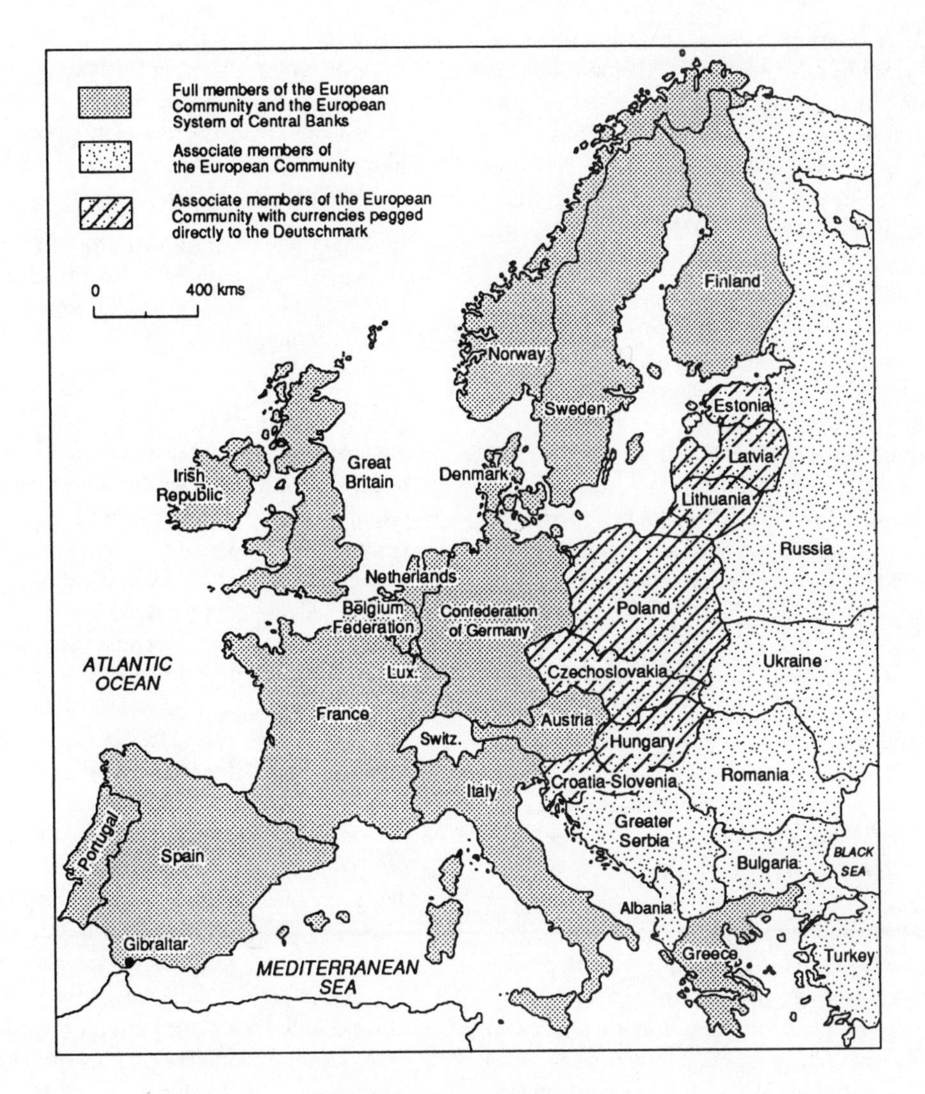

Figure 3.5 Europe in 2000 AD?

controlling coal and steel production in the member countries.

The most important move towards European integration came with the establishment of the EEC in 1958, after the Treaty of Rome was signed in 1957. The EEC was composed of the six countries of the ECSC. It was designed to allow free trade in a common market and establish joint economic policies. The common market would allow the free movement of people, capital and goods. A tariff-free market was set up within the EEC and common tariffs were extended to the outside world. The two strongest economies in the EEC were West Germany and France. The success of the EEC depended upon their compromises. The Germans, with their strong industry, wanted a tariff-free market to aid the export of manufactured goods; the French, with their large, still essentially peasant agricultural sector, wanted protection for agriculture. The result was a fiscal trading system designed for economies with strong industry and weak agriculture. In later years Britain was to suffer from this arrangement since she had a strong agricultural sector and a weak industrial base. The original signatories strengthened their links in 1967 by setting up the European Community, which brought together the EEC, the ECSC and the European Atomic Energy Commission.

The EEC was becoming one of the biggest markets in the world and non-EEC countries in Europe were eager to join. Britain tried to join in 1963 and again in 1967, but both times British entry was vetoed by de Gaulle. The French President's arguments were simple: Britain was tied too closely to the USA and Britain's entry, according to de Gaulle, would increase US influence within the EEC and vitiate the importance of France. Britain's application to join the Community in 1972 was more successful. In 1973 the original Six expanded to the Nine with the entry of Britain, Eire and Denmark. Greece joined in 1981 and Spain and Portugal in 1986. In 1992 a single market was created, and with it an economic power to rival the USA and Japan.

The dominant economic theme underlying policies in the European Community has been the belief in a free internal market, with a minimum of restrictions, to allow the unhampered movement of goods, people and capital. This belief is based on the assumption that, holding everything else constant, a lowering of tariffs and a reduction of restrictions would stimulate economic growth. And so it has. The problem for European integration has been that this growth has been uneven within and between member countries. There is a definite core–periphery structure to the European Community which creates disparities in economic growth, income and overall quality of life. There is now a 600 per cent difference between average incomes in the western part of the newly united Germany and those in southern Italy.

Such regional and national disparities hinder European integration. Countries which lose out in the process of uneven growth can become centres of discontent, urging protectionism and threatening the political stability of the Community. Depressed areas with the potential for political mobilization (e.g. Scotland, Wales, Corsica, Brittany) can also become the scene for secessionist movements of varying degrees of intensity.

The two most important external issues affecting the European Community are:

- the economic position of European capital in an increasingly competitive world market
- its diplomatic and strategic relations in a multipolar world.

1 The European Community has aided the growth of large European companies, whose ownership resides in Europe although they may operate around the world. The Community provides a large home market for companies (part of the reason for its creation was to provide such conditions) and the Community has explicitly aided mergers between European firms. Community policy directives have been an important, although not the sole, reason for European mergers between, for example, Dunlop–Pirelli and Agfa–Gavaert. The large home market facilitates the trend towards larger European companies and these companies can now mobilize the vast amounts of capital necessary to invest overseas and in research and development. European companies can now compete successfully with North American and Japanese firms throughout the world. This means that political alliances between the USA, Europe and Japan are becoming strained by their economic competition for world markets.

2 One element in the break-up of the bipolar structure has been the emergence of a European stance on world affairs as distinct from that of the USA. There has not been a coherent or even an agreed European foreign policy, and many would argue that the differences between the European nations are as great, if not greater, than the differences between individual European countries and the USA. However, what seems clear and irrefutable is the gradual emergence of a world view held by different European countries to varying extents, but held nevertheless, which differs from that of the USA. The

world views overlap substantially in places but the measure of disagreement is apparent and in certain areas quite large.

The New Europe

The 1990s saw two significant changes in European affairs;

1. The single market of 1992, when the EC became a united single market free from internal tariffs of member states. Internationally, this means the EC will be even more of a single economic unit in competition with the USA and Japan. Internally, the market may lead to the creation of a much sharper core–periphery structure.
2. The decline of communist party power in Eastern Europe. Germany has already been united and future developments may include an expanded EC in which Poland, Czechoslovakia, Romania and Bulgaria are members; maybe even the countries which until very recently made up the Soviet Union. Figure 3.5 suggests one possible scenario for Europe by the end of this century.

This is an interesting time in world affairs. Old structures are falling around our ears, old ideologies are evaporating and new relationships are emerging. Who could have predicted the events of 1989 or the break-up of the USSR in 1991? It is difficult to forecast the course of future development; what is evident, however, is that the old bipolar world order is over, and one of the most important elements of change is the emergence of an integrated Europe.

GUIDE TO FURTHER READING

On China have a look at:

Dreyer, J. T. (ed) (1989) *Chinese Defense and Foreign Policy.* Paragon, New York.

Gittings, J. (1990) *China Changes Face: The Road From Revolution, 1949–1989.* Oxford University Press, Oxford.

Gray, J. (1990) *Rebellions and Revolutions.* Oxford University Press, Oxford.

Harding, H. (1987) *China's Second Revolution.* Brookings Institution, Washington.

Lu, L. (1990) *Moving The Mountain.* Macmillan, London.

Mancall, M. (1984) *China At the Center: 300 Years of Foreign Policy.* Free Press, New York.

Schell, O. (1988) *Discos and Democracy: China in the Throes of Reform.* Pantheon, New York.

Spence, J. (1990) *The Search For Modern China.* Hutchinson, London.

On Japan the following is just a sample from a wide literature:

Drifte, R. (1990) *Japan's Foreign Policy.* Routledge, London.

Duus, P. (ed) (1990) *The Cambridge History of Japan: Volume 6.* Cambridge University Press, Cambridge.

Ecclestan, B. (1989) *State and Society in Post-War Japan.* Polity Press, Cambridge.

Hendry, T. (1989) *Understanding Japanese Society.* Routledge, London.

Lincol, E. J. (1988) *Japan: Facing Economic Maturity.* Brookings Institution, Washington.

Reischauer, E. O. (1988) *The Japanese Today: Continuity and Change.* Harvard University Press, Cambridge, Mass.

Suzuki, Y. (1987) (ed) *The Japanese Financial System.* Oxford University Press, Oxford.

Tasker, P. (1987) *Inside Japan.* Sidgwick & Jackson, London.

Yamamura, K. and Yosuba, Y. (eds) (1987) *The Political Economy of Japan.* Stanford University Press, Stanford.

For Europe, sample some of the following:

Blacksell, M. (1977) *Post-War Europe: A Political Geography.* Dawson, Folkestone.

Brown, J. F. (1988) *Eastern Europe and Communist Rule.* Duke University Press, London.

Clout, H. (ed) (1987) *Regional Development in Western Europe.* Fulton, London.

Clout, H. *et al.* (1989) *Western Europe: Geographical Perspectives.* Longman, London.

Palmer, J. (1988) *Europe Without America: The Crisis in Atlantic Relations.* Oxford University Press, Oxford.

Palmer, J. (1989) *1992 and Beyond.* Office for Official Publications of the European Communities, Luxembourg.

Pinder, D. (ed) (1990) *Western Europe: Challenge and Change.* Belhaven Press, London.

Urwin, D. W. and Paterson, W. E. (eds) (1990) *Politics in Western Europe Today.* Longman, London.

Relevant journals

East European Quarterly
European Affairs
European Journal of Political Research
Journal of Asian Studies
Journal of Contemporary Asia
Journal of Japanese Studies
Pacific Affairs
West European Politics

Other works cited in this chapter

Kahn, H. (1971) *The Emerging Japanese Superstate.* André Deutsch, London.

Powell, E. (1988) 'The decline of America', *The Guardian*, 7 December p. 23.

Short, J.R. (1982) *An Introduction to Political Geography.* Routledge & Kegan Paul, London.

Vogel, E. (1979) *Japan as Number One.* Harvard University Press, Cambridge, Mass.

PART II

THE POLITICAL GEOGRAPHY OF THE STATE

The building block of the world political order is the state. For a long time the state was a neglected topic in political geography. It was often incorrectly used as a substitute for 'nation'. Too often it was treated as the repository of a vaguely defined general interest, a neutral tool, an innocent variable. Recent work has challenged this cosy view. The state itself is now seen as an arena for competing interests. The next part of this book concentrates on the state. Chapter 4 looks at the range of states, Chapter 5 considers the tensions implied in the notion of the nation-state, while Chapter 6 examines some of the implications of the state as a spatial entity.

4

THE STATE AND THE WORLD ORDER

The policy of a state lies in its geography.
(Napoleon Bonaparte)

One of the most important developments of the twentieth century is the growth of the state. Growth in the double sense of:

- There are now more states than ever before. In 1930 there were only about 70 sovereign states. By 1990 this figure had grown to over 160.
- In most countries the state has become bigger and more powerful.

The state has a number of goals and functions. Table 4.1 notes some of the more important. The ability to pursue these goals depends upon the position of the state in the political and economic world order. Let us consider the relationship between states and the world order.

Table 4.1 Functions of the state

Functions	*State apparatus*
External	
Defend the country	Diplomatic service, armed services
Maintain favourable trading relations	Diplomatic services, armed services, government agencies which sign trading agreements, give aid to domestic producers, create tariffs to keep out foreign competitors
Internal	
Maintain law and order	Police force, armed services, judiciary
Maintain belief system	Education and mass media (propagating essential rightness of existing system)

THE STATE AND THE WORLD POLITICAL ORDER

States are neither equal nor similar. States vary in power and influence. As a rough guide, we can distinguish between superpowers, major powers and minor powers.

Superpowers are those countries which have a global capacity to influence events. For them, the world's surface is like a giant chess board where the moves and counter-moves indicate the changing balance of power. *Major powers* are the knights, bishops and rooks of the chess analogy. They can be superpowers on the way down (e.g. present-day Britain) or minor powers on the way up (e.g. Japan). Although they have less global influence than the superpowers they have either strategic or commercial interests throughout large parts of the world. Japan, for example, has trading links throughout the world though its present lack of military capacity currently denies it superpower status, while China has an important diplomatic and strategic role in its relations with the USA and Russia but lacks the economic and military capacity at the moment to be more than a major power. *Minor powers* have a very limited direct role or influence in world affairs. They are the pawns of the chess board. But just as pawn swaps can determine a game's outcome, so the relations and conflicts between minor powers in areas of strategic importance, the 'fracture lines' of geopolitical significance, can take on much wider importance. The position of minor powers in sensitive areas of geopolitical space (e.g. Poland and Afghanistan, Lebanon and Nicaragua) can raise their internal disputes to global significance.

The ranking of countries in this three-fold division is not a permanent categorization. A shift in category reflects a country's changing geopolitical role. Table 4.2 provides a very brief selection of some of the changes.

Table 4.2 Changing geopolitical status, 1890–1990

	1890	*1990*
Superpower	UK	USA, USSR
Major power	USA, Russia	UK, Japan, China
Minor power	Japan, China	Poland, El Salvador

Let us look at these categories in more detail.

Superpowers

The goal of superpowers is to maintain their position on top of the world order. They achieve their status through economic power, military might and political influence. Their aim is to incorporate as much of the world as possible into their sphere of influence. *Spheres of influence* can be defined as the territory in which a superpower can wield effective power. What makes a superpower is the size of the sphere and the degree of influence. Influence is the ability to reach a desirable outcome. For example, the USA does not want to see a socialist state established in the Western Hemisphere. With the notable exception of Cuba it has been very successful in this goal. This has been achieved by diplomacy, flexing of economic muscles (e.g. refusing aid or trade, imposing embargoes) and military involvement, either direct or indirect. After the leftist revolution in Nicaragua in 1979, for example, the US government stopped aid to that country in 1981, imposed a trade embargo in 1985 and aided anti-government forces, the *contras*.

The ultimate power of a superpower is to influence events without even trying. In Central and Latin America, for example, all politicians know that the establishment of a socialist society would be frowned on by the USA, which would exercise its power to stifle the change. This is so much part of the political reality that it often militates against such change.

The sphere of influence of a superpower does not stretch evenly throughout the world. Most superpowers have a home base, which could cover either all or only a part of the state territory. In the USA it covers all the country, but in the USSR it was centred on the ethnic Russians. Then there are core area(s) of control in which one superpower reigns supreme and, finally, peripheral zones of competing power.

Superpowers attempt to maintain and improve their geopolitical position by extending their relative spheres of influence. This is achieved through maintaining control in their respective cores – for the USA this extends southward into Central and South America, northward into Canada and eastward across the North Atlantic into Western Europe. In contrast, the USSR's core used to include most of Eastern Europe. Within these relatively long-established areas of control the role of the other superpower has been limited, with the Monroe and Brezhnev Doctrines being examples of 'hands-off' notices. In these areas the dominant superpower has maintained control through traditional diplomacy, direct intervention (e.g Grenada in 1983 and Czechoslovakia in 1968) and indirect intervention (e.g. El Salvador and Poland in the early 1980s).

Outside of these core areas the rest of the world constitutes a periphery of competition, including less well-established spheres of influence, neutral areas and zones in flux where the mediating elites are either changing alliances or new elites are emerging. Zones of instability can be defined in relation to the posture, cohesion and stability of the mediating elites. It is these areas of the world, particularly in Eastern Europe, the Middle East

Table 4.3 Changing spheres of influence since 1960

USA→USSR	USSR→USA	USSR→neutral	USA→neutral
e.g.	e.g.	e.g.	e.g.
South Vietnam	Egypt	Yugoslavia	Iran

and South-East Asia, which sometimes give the superpowers the opportunity to incorporate a country into their sphere of influence and which are the scenes of superpower rivalry – rarely in terms of direct confrontation but more commonly in terms of conflict through backing warring factions. In these unstable zones, the fracture lines of the world, the position of individual countries in relation to the superpowers can change, either as established elites change alliances or a newly established elite changes the country's stance. By way of example, Table 4.3 indicates various shifts in spheres of influence.

The categories have to be used with care since there are few countries which took either a totally pro-USA or pro-USSR stance in all matters, or a completely neutral position. The category 'USA USSR' implied a shift along a continuum rather than a change from one state to another. With this in mind, Table 4.3 gives one example of each of the categories in the 1945–90 period.

If we look through the historical record we find that all superpowers have two characteristics:

- they legitimate their power with an ideology;
- they don't last forever.

Ideologies of empire

The actions of superpowers are codified in geopolitical strategies which try to legitimate their international role. In the sixteenth and seventeenth centuries Spain saw itself as the bearer of Christian virtue, a beacon of Catholic light in a pagan darkness. In the nineteenth century the British Empire clothed its world role in terms of a civilizing mission. By the mid-twentieth century the USA saw itself as the world's policeman saving countries from the evils of communism, while the USSR had the ideology of Marxism-Leninism to give credence to its self-image of torch-bearer for world socialism.

These ideologies did not go uncontested. Criticisms of the imperial posture can always be found (see references to Hobson and Williams in chapters 1 and 2) but they remain dominant as long as the country remains a superpower. The ideologies give an intellectual rationale and an ideological justification to the more mundane commercial and strategic concerns of a superpower.

Imperial overstretch

Superpowers do not last forever. Their rise and fall has been described by two scholars Modelski and Kennedy. Modelski (1983, 1987) suggests a series of long cycles from 1494 to the present day. Each long cycle consists of three phases:

1. From a generation of global war one state emerges strongest. This state becomes a superpower able to arrange the global political and economic order to its advantage.
2. Challenges to this order are made by other states which have been growing in power. The world becomes more multipolar, even though one state may still be dominant.
3. The lull before the storm: the dominant superpower begins to weaken and challengers gain confidence before mounting their claim for dominance. Then we are back to phase one, as in the case of Britain in the early nineteenth century after the Napoleonic Wars or the USA after the Second World War.

The three phase model is summarized by Modelski as: one generation builds, the next consolidates, and the third loses control.

In a book entitled *The Rise and Fall of the Great Powers* (1987) the British historian Paul Kennedy tackles the question of why superpowers decline. His answer is to point to the phenomenon of *imperial overstretch*. Superpowers arise on the basis of their economic and military strength. But it becomes more and more difficult to maintain this position especially in phases two and three of the

Modelski cycle. More and more of their expenditure is devoted to maintaining their power against enemies, real and imagined; the 'frontiers of insecurity' have to be defended. Eventually these imperial commitments undercut the economic strengths of the state. The initial basis of power is weakened. In 1970, two years before Henry Kissinger dates the end of the bipolar world, the USA had one million soldiers in 30 countries, 5 regional alliances, 42 defence treaties, membership of 53 international bodies and provided economic aid to over 100 countries. Now that's what most people would call overstretch!

Kennedy provides examples of powers at their imperial peak – Spain in 1600; Bourbon and Bonapartist France; the Dutch Republic in 1600; Britain in 1860 – and shows how overstretch got them in the end. The book was an unexpected success, selling over a hundred thousand copies in its first year of publication in the USA. Its success is no doubt related to the fact that it could be addressing the present condition of the USA.

Major powers

There is a continuum from a dominant superpower to a minor power. The major powers fall between these two extremes. They consist of former superpowers on the way down (e.g. Britain) and former minor powers on the way up (e.g. Japan). Former superpowers have a legacy of empire. Let us elaborate this point with respect to Britain, *the* superpower of the nineteenth century, with enormous territories (see Figure 4.1).

The experience of empire aided the accumulation of wealth in Britain. This wealth was not evenly distributed, it was concentrated in a few hands. Nevertheless, much of the population gained as the wealth enabled relatively high wages to be paid. The accumulated wealth also took the form of art treasures, cultural artefacts and the like. Britain has some of the best museums in the world as a result of imperial plunder. Kew Gardens and the British Museum house the booty of the British Empire as much as the Bank of England.

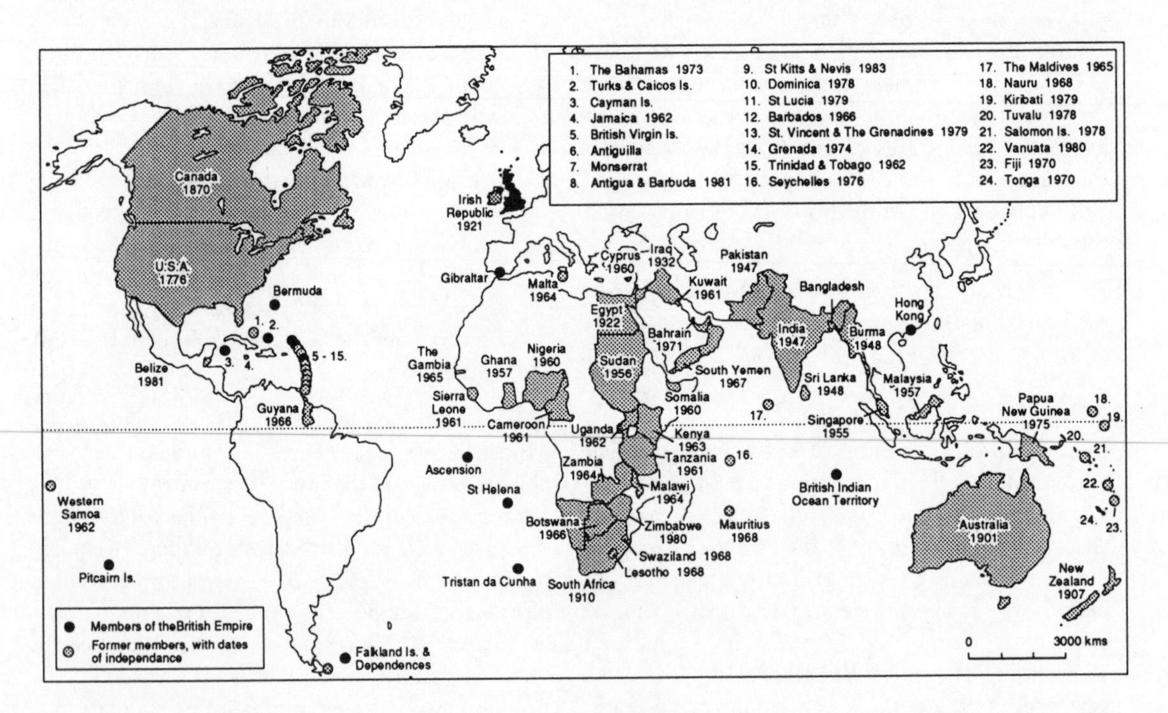

Figure 4.1 The British Empire

A major imperial legacy was an open market, i.e. one dependent on high levels of import and export. This type of economy is successful with competitive home producers and secure export markets. However, with the loss of these markets in the wake of decolonization and decline in the productivity of home-based producers, the open market has become a scene for high import penetration and decline of domestic production.

The experience of empire also impinged on ideas. We can identify three separate elements:

- The success of the nineteenth-century informal empire gave extra purchase to free trade ideologies in Britain. In Manchester's Free Trade Hall, Britain has one of the very few buildings dedicated to an economic doctrine. The ideology of free trade is the voice of the economically successful who want open competition. However, the ideology has persisted even though Britain's relative economic performance has declined. All countries remember the time when they were most successful; the ideologies of that period persist.
- The annexation of territories involved contact with other racial groups. But contact of a particular type. The black and brown colonial peoples were often perceived as inferior. With the immigration of some Asians and West Indians in the 1950s and 1960s in response to job opportunities and promotion by British firms (in the 1950s, for example, London Transport recruited labour directly in the West Indies), these racial stereotypes were sometimes employed to stigmatize the ethnic minorities. The race 'problem' in Britain is primarily one of the attitudes initially formed and strengthened in the colonial experience.
- In the nineteenth century Britain was the superpower, both in commercial and military terms. A world role for Britain was seen as 'legitimate', 'natural' and 'necessary'. These attitudes persisted throughout the twentieth century. It is only comparatively recently that they have changed. The military presence east of Suez, for example, was not an election issue in the 1950s. However, the officially perceived need for a world role persisted and was evident in Britain's substantial military spending, its vast and expensive diplomatic service and its independent nuclear capability, which owes more to officially perceived national self-esteem than to rational strategic considerations.

There are also those physical fragments of empire scattered about the globe which for one reason or another, the precise combination varying by individual case, remain under the British flag (see Figure 4.1). These range from the Falklands to Gibraltar and Hong Kong. The Falklands War of 1982 between Britain and Argentina is a reminder of the extent to which old colonial annexations could still be a source of armed conflict. Existing colonial holdings are also the cause of delicate negotiations: with China in the case of Hong Kong and Spain over the Gibraltar issue. The last withdrawals from empire are likely to be as hazardous as the earlier ones.

Britain's former imperial power still has global consequences. One of the most important is

Figure 4.2 The world centred on London

language. Whenever Britons took control they established their language as well as their political and economic power. Throughout the world English was spoken and even today it is one of the world languages, almost *the* world language.

The way the world is described is also a function of Britain's position (see Figure 4.2). From London, Egypt is in the *Middle East* and Hong Kong in the *Far East*, Vietnam is *South-East* Asia. These terms are still used today. We take such terms so much for granted that we forget it is a convention related to Britain's imperial position. Think of an alternative. If the empire had been centred on Jerusalem, Britain would be the Middle West, the USA would be the Far West and Chile would be in South-West America. In the case of Japan we can see a superpower in the making. Since the mid-1960s Japan has emerged as one of the strongest economies in the world. Hard work, efficient work practices and the free trade arrangements of its political allies have enabled Japanese industries to penetrate the large markets of Europe and America. Japan is in the position of challenging the global supremacy of the USA unhindered by imperial overstretch. Japan has sheltered under a defence umbrella provided by the USA. How long this can continue is a moot point. The USA wants Japan to shoulder more of the defence burden. Moreover, to sustain its economic dominance Japan may have to enter into direct military arrangements.

We should be careful of reading domestic conditions from a state's world position. A waxing superpower is not necessarily a better place to live for the majority of the population than a waning superpower. The lot of the average citizen in Britain is much better in the 1990s than ever it was in the 1890s when the British Empire was at its greatest extent. In a waning power there is less need to spend money on maintaining a world position, more can be spent on provision of services which improve the quality of the citizens' lives. We can change Modelski's phase: one generation works hard, the next generation works, and the next may begin to enjoy life.

In specific regional contexts major powers can exert an influence. Their sphere of influence may be smaller than a superpower's but still be very strong. Former superpowers may have a sphere of influence in areas of the world where they had a colonial presence. France, for example, has an influence with former colonies in Africa and a continuing hold on territories in the Pacific. Britain is another example of a major power in which the legacy of empire extends to present day influence on former colonies. This influence can be through the informal ties of shared language, similar institutions, the education of elites, or more formally in such organizations as the British Commonwealth.

Minor powers

Most countries in the world are minor powers of varying degrees of importance. In 1988, of the 159 members of the UN, less than 10 had nuclear weapons and most of them have an economy smaller than the wage bill of IBM or General Motors.

We can identify different types of minor powers. There are those which are *minor but not poor*. Switzerland, for example, has one of the highest standards of living in the world but plays very little part in the world political order. It remained neutral during the two world wars of this century and has retained its independence from a number of international organizations; despite being in Europe it is not part of NATO or the EC. Then there are the *poor and minor* countries such as Bangladesh. In the normal working of the political order the minor powers are caught up in the ebb and flow of superpower manipulation or major power influence.

Although there have been examples where countries have broken away from the spheres of influence, e.g. Yugoslavia and Iran, minor powers usually find it very difficult to maintain an independent position. They need support from other countries, in the form of aid, in regional alliances, or in world organizations. One reason behind the growth of international forums (e.g. the United Nations) is that they give a voice to the small countries. Large international forums are strongly supported by the minor powers because they allow them a say, and act as a counterweight to the concentrated power of the superpowers and major powers.

THE STATE AND THE WORLD ECONOMIC ORDER

The ability of a state to pursue the goals identified in Table 4.1 also varies according to its position in the world economic order. At the risk of over-simplification we can identify three different types of state:

- the state in advanced capitalist countries;
- the state in the semi-periphery and peripheral areas;
- the state in socialist countries.

The state in advanced capitalist countries

In chapter 1 it was shown that the core countries of Western Europe, Japan and North America had benefited from the structure of the world economy. The flow of wealth from the semi-periphery and periphery to the core, has meant that labour has managed to obtain relatively high wages while government revenues have been increased through a broad and deep base of taxation. The result in the post-war era has been rising living standards and a measure of social harmony. The lack of revolutionary upheaval in the core has been partly due to the ability of the economy to create and meet rising expectations. For countries firmly established in the core, economic growth continues to maintain the basis for social peace. When the economy is booming there is no problem. Tensions arise when the national economy is facing competition. The state becomes the arena for the ensuing conflict. It is called upon by threatened domestic producers and labour to introduce some form of import controls but business interests which are relatively successful in the export market continue to preach the benefits of free trade.

We can identify a number of different perspec-

(a) Democratic-pluralist

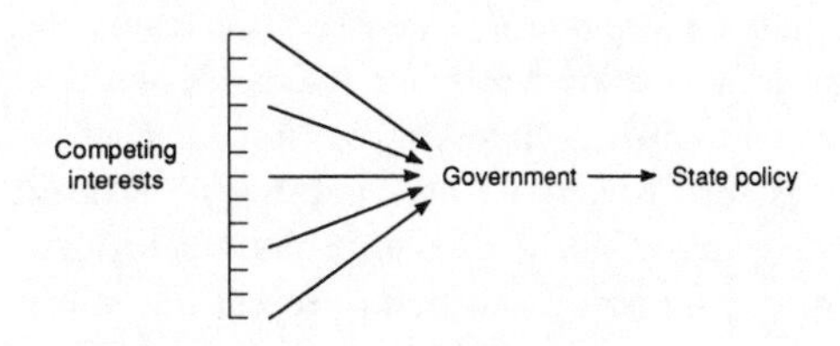

(b) Imperfect-pluralist

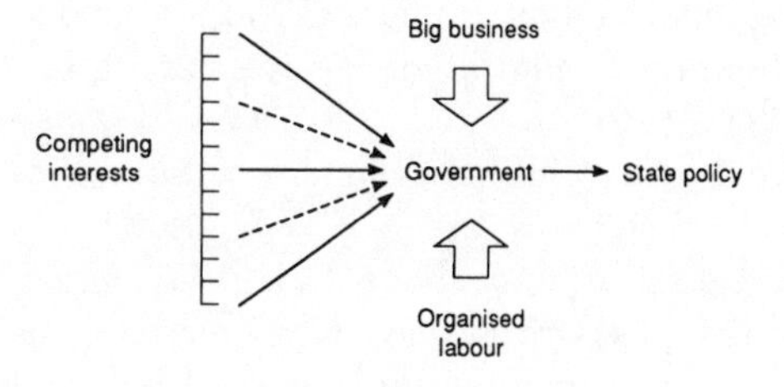

(c) Marxist: instrumental

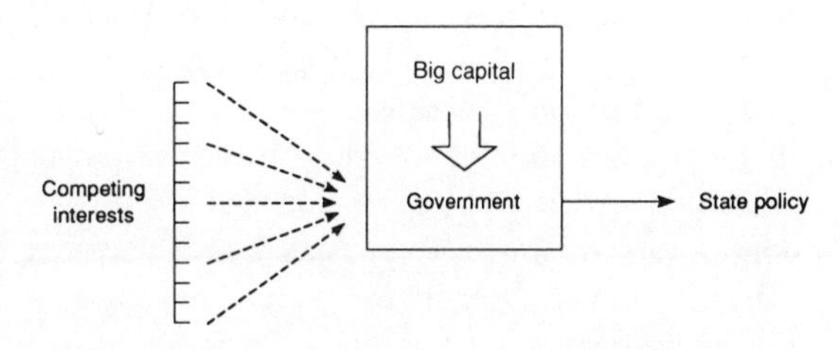

(d) Marxist: balance of class forces

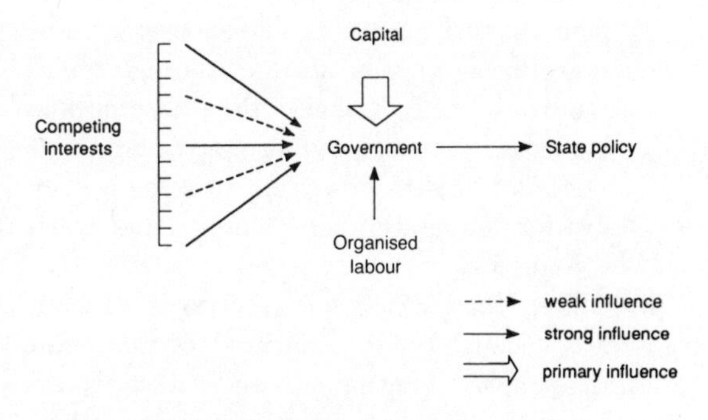

Figure 4.3 Alternative models of state policy

tives on the role and functioning of the state in core countries.

The *democratic-pluralist* view sees the state as a neutral arbiter, a kind of giant referee between a number of competing interests. In this perspective society is seen to consist of a number of diverse pressure groups. These organizations are formed around specific issues and/or to protect the interests of particular groups. They achieve, or try to achieve, their goals through formal and informal politicking. The democratic-pluralist position asserts that competition between the groups, their diverse representation and the sheer variety tend to cancel out the dominance of any one social group (see Figure 4.3a).

The initial democratic-pluralist position, as mapped out by writers such as Dahl (1982), has been subject to an autocritique (Dahl, 1989) and to revision by those writers who criticize the emphasis on issues and competing groups (Bachrach and Baratz, 1963). The crux of the criticisms is that power can reside in the stages prior to overt decision-making. Power can be exercised so that only relatively safe issues appear on the political agenda. Certain powerful groups can exercise power so that only those issues which involve no major redistribution of wealth, privilege or power can come up for public discussion and debate. To concentrate on observable issues and interested groups would be to ignore the exercise of power on the selection of issues for debate. In this modified version of the democratic-pluralist model there would seem to be a two-stage process:

(a) The political agenda is shaped by the more powerful groups in society, through direct lobbying or through the assumptions embedded in a common ideology shared with the political elite. Power at this stage is exercised at the level of policy choices and policy formulation.
(b) Those issues which surface onto the agenda are decided by the outcome of the more observable bargaining and competition between the various interest groups. Here, power is restricted to influencing policy implementation. The state is the agent which mediates between the two stages.

The modified model can be termed *imperfect pluralism* to refer to the fact that there are a variety of interest groups but some are more important and successful than others. This line of thought sees public policy not as the result of numerous interest groups and atomistic voters, rather it is dominated by the big battalions of organized labour and the dominant factions of capital (see Figure 4.3b). The argument is that the democratic-pluralist state is, in reality, more of a corporate state.

The *Marxist* view constitutes the major alternative perspective on the state. There is no fully developed Marxist theory of the state; rather there is a continuing debate between various schools of thought. However, they all share the same general view which repudiates the notion of the state as a neutral arbiter. The state is not above the class struggle and social conflict in society, it is part of them. Marxists view all societies as class societies in which one class dominates the others and the state is the means and support of this class domination. The state acts on behalf of the ruling class but in capitalist society, particularly advanced capitalist societies, it has a degree of relative autonomy from all classes. This relative autonomy is the 'distance' between the character of the different classes and the form of the state; it is the degree of freedom which the state has from immediate class interests in determining policy. Within this broad consensus different approaches can be identified. Let me note just four examples (see Table 4.4).

Table 4.4 Approaches to the state

Dominant function	*Degree of relative autonomy*	
	Greater	*Lesser*
Economic	Responses to crises	Underconsumption
Political	Balance of class forces	Instrumentalism

See text for discussion of terms.

In concentrating on the political function of the state two contrasting models can be identified. The *instrumentalist* approach, associated with the early work of Ralph Miliband (1969), seeks to show how the state is an instrument of the ruling class. A common method of research is to show the common background and shared life experience of the capitalist class and important state functionaries and officials. Miliband's work in Britain, for example, shows how senior politicians and important civil servants have similar backgrounds to industrialists and large capital investors. The conclusion drawn is that this shared background leads to common aims, assumptions and ideologies; the net result is for the general direction of state policy to be in the interests of the capitalist class (see Figure 4.3c).

The *balance of class forces* model takes a more complex view of things. In this model the state is an arena for competing interests and not simply the vehicle for class domination. The state represents the power of the capitalist class, but there are various fractions of this class with different interests. To achieve social stability the state will often sacrifice the interests of the weaker fraction to meet the expressed needs of the population. The dominant fraction is big business. Small businesses have little say in the development of state policy.

According to this view the state is not a monolithic apparatus always reflecting the interests of (big) capital but the scene of social conflict and compromise; state policy reflects and represents the broad balance of political power between different social groups and classes (see Figure 4.3d).

In terms of the economic function of the state two contrasting models can be identified. The *underconsumptionist* school of thought has been the predominant Marxist interpretation of the state in the USA and is associated with the influential work of Baran and Sweezy (1960). Baran and Sweezy view the state primarily as a mechanism for maintaining effective demand and ensuring 'surplus absorption'. In their analysis of the social and economic order in the USA, they argue that under oligopolistic conditions the economic surplus tends to rise. The surplus is defined as the difference between what a society produces and the socially necessary costs of production. It tends to rise because the large firms collude to maintain the stability of prices even after the introduction of cost-saving improvements. The oligopolistic quality of the market ensures that competition does not lead to price reductions and there is a tendency for stagnation as the productive capacity for consumption goods expands faster than effective demand. Two mechanisms have arisen to solve this realization problem:

- the sales effort, which promotes continual consumption
- the state, through high levels of state expenditure, which includes non-defence purchases (e.g. education), transfer payments (e.g. welfare payments) and defence spending.

The Baran and Sweezy model is very specific. A more general model is *response to crises*. This is a model which suggests that one of the major tasks of the state in capitalist society is to manage the crises which are inherent in the capitalist mode of production. Crises can take a number of forms: falling rates of profit, rising unemployment, rapid inflation. The crises become a political phenomenon because the state is an arena for the competing interests, and crises can only be solved with major redistributional consequences. The major cleavage is between capital and labour. The nature of the economy ensures the need for continued state intervention while the exact nature of the intervention depends upon the balance of class forces.

In recent years there has been a move in Marxist theories from specific models which read off the interest of capital from state policy to looser descriptions which give more leeway to the precise configuration of social forces. There has been a move, when looking at the political role of the state, from instrumentalism to balance of class forces and, when looking at the economic role, from underconsumptionism to response to crises. This move is paralleled in non-Marxist theories, from democratic-pluralism to imperfect pluralism.

There is a growing convergence between Marxist and non-Marxist views. To summarize the area of agreement between them:

- there is a belief that the state is an arena for competition between different social groups
- some of these groups are more powerful than others
- the most powerful are big business
- in the long term the state seems to reflect the interests of big business more than that of small business or labour
- this does not mean that the state does not respond to democratic pressure in order to achieve social stability
- the state is subject to popular pressure and state policy at any one time reflects the balance of power between the competing groups.

It seems as if the democratic models are becoming less pluralist while the Marxist models are becoming less Marxist.

The state in the capitalist periphery

There are very large differences between the form of the state in different peripheral and semi-peripheral countries. In this section the comments will therefore be of a general nature.

The overall form of the state in peripheral capitalist countries has been shaped by:

- the incorporation of that society (and its predecessors) into the capitalist world economy
- the changing nature of the economic relations with the core.

Each of these elements can be examined separately, although in reality they are resolutely interwoven.

Incorporation

The nature of incorporation has done much to shape the subsequent development of peripheral societies. For those countries in Africa and Asia, for example, which were incorporated during the age of imperialism, the post-colonial society has inherited a powerful state with a large bureaucratic (and often military) apparatus which has been used to regulate and control large areas of social life. The state was, and continues to be, the linchpin of economic, political and social development, and control of the state apparatus becomes the prime consideration for political movements and social groups. The state has taken on even greater significance in those countries which lack a national history. In many African countries state boundaries often reflect cartographic convenience or the legacy of imperialspheres of influence rather than any logic of national identity. In these circumstances the state is used to create the experience of nationhood.

In contrast to the core countries, where the separation of economic and political spheres is held to exist (especially by right-wing politicians) either in the past or the present, societies in the periphery inherited a state apparatus designed to order economic development and to control the direction and flow of trade. In post-colonial societies the state is inextricably linked with the economic realm. The state in the periphery thus becomes a source of economic power, and private wealth can accrue from political power to an even greater extent than in the core.

The nature of political developments in post-colonial societies has also been guided by the process of de-colonization. The overthrow of foreign political control favoured the emergence of a dominant single party or mass movement. In the conditions of a political and sometimes military struggle, the emphasis was placed on one mass movement or a single political party and strong internal discipline. These features often continue in the post-colonial era in the form of one-party states.

Changing economic relations with the core

The character of the peripheral state can also be seen against the background of changing economic relations with the core. Let us consider just one area of the world. In Latin America, for example, a number of social groups can be identified:

- a powerful class of landowners whose interests are tied closely to those of foreign capital
- an indigenous bourgeoisie whose interests lie in stimulating autocentric economic growth
- the vast mass of the population who are either urban dwellers and workers or peasants.

The history of the state and state policies in Latin America can be seen as the evolution of conflict and alliances between these different groups against the backcloth of external links with the core. Munck (1979) suggests a three-phase sequence:

1. the export phase up to 1930, when the free-trade parties representing the landowners were in power and the state was involved in providing the necessary infrastructure facilities for an export-based economy. The job of the state was to facilitate this relationship by providing the necessary infrastructure of roads, ports and railways and by providing, maintaining and controlling a suitable labour force.
2. The transition to an industrial state from 1929 to the 1950s, when economic links with the core were weakened by world war and economic depression. The economic conditions allowed the possibilities for autocentric economic growth and the emergence of a much stronger indigenous bourgeoisie. The state's role was expanded from that of simply providing an infrastructure to providing the fiscal and legal framework for industrialization. The transition period in many countries saw the growth of an urban working class and a more organized peasantry. These developments saw the introduction of a new political force of explicitly socialist persuasion.
3. From the 1950s to the present day is the phase of industrial growth. The growing power of the indigenous bourgeoisie has forced the state to expand its role from providing the conditions for capitalist production to being directly involved in the production and exchange of commodities. The state has become involved in those large, lumpy investments which private local capital cannot find and which provide too low returns for foreign capital. The state is also involved in direct investment. The exemplar case is Brazil, where the state has undertaken massive direct investment.

In the periphery and semi-periphery the state's function is to attract, guide and control foreign investment, to stimulate internal economic growth and to maintain the impetus for change in the economy through various international organizations and commodity cartels. In each of these areas the balance of power has shifted from foreign capital to the state, with the degree of shift in each country reflecting the size and strength of the economy, the amount of natural resources within the territory and the political and social character of the ruling elite.

The state in socialist countries

There were substantial differences between the form of the state in socialist countries in such diverse places as Vietnam, Cuba, Poland and the Soviet Union. No brief review can hope to describe, let alone explain, this variety; the aim will therefore be to describe those very general features of the state which were markedly different from those in capitalist countries.

Successful revolutionary socialist movements obtained power:

- in peripheral economies
- within the context of a hostile world
- led by political parties armed with a comprehensive and encompassing ideology.

Each of these three characteristics influenced the form and role of the state.

Peripheral economies

It is one of the political facts of this century that socialist movements with an explicit Marxist ideology have succeeded in obtaining power only in more peripheral countries of the world. Thus the early emphasis of their political leaders was on stimulating economic growth and development. There was a real and perceived need to 'catch up' with the core countries. Socialist states have attempted to telescope the long historical experience of the core into a series of rapid programmes of socialist economic growth. The state has been afforded a huge role in this process. Under centrally planned economies the state is the key unit of economic decision-making; it makes priorities and sets targets:

> According to the logic of the centralized planning model, the central organs know what society needs, and can enforce plan-orders to ensure that these needs are effectively and efficiently met. This requires multi-million instructions as to what to produce, to whom deliveries should be made, and from whom inputs should be received and when. All this must be made to cohere with plans for labor, wages, profits, investment financing, material-utilization norms, quality, productivity, for each of many thousands of productive units.
>
> (Nove, 1980, p. 4.)

This planning system has certain advantages and disadvantages compared with the capitalist system. In drawing up a balance sheet of Soviet economic performance, Nove (1980) points out the high degree to which definite priorities can be set in a planned environment, the low levels of unemployment and the stability of wages and prices. But the system had its deficiencies. Centralized planning in the USSR was unwieldy and could not cope with sudden changes in demand, and innovation was stifled. Bottle-necks frequently occurred as outputs from one sector failed to mesh with the needs of other sectors and many poor-quality goods were produced.

Hostile world

Revolutionary socialist movements have overthrown existing social and economic orders in the context of a hostile world. The socialist state bore the imprint of the revolutionary experience in the role of the party. Revolutions are made by the masses but are led by political parties. In *What Is to Be Done* (1902), Lenin argued that revolutionary potential could only be realized and revolutionary advance could only be possible through a highly organized, close-knit, strongly disciplined party. The party gave the revolutionary process direction, organization, strength and success because 'Without a guiding organization the energy of the masses would dissipate like steam not enclosed in a piston-box' (Trotsky, 1977 edn, p. 19). The highly centralized nature of the party, understandable in a revolutionary period, tended to continue in those countries where the Leninist line was followed. Bureaucratic inertia, unwillingness to give up power and the fear of external attack and internal sabotage all combined to concentrate power in the post-revolution period. The dangers in such a situation are obvious. Trotsky's remarkably prescient comments of 1903 became chillingly true in Russia under Stalin:

> Lenin's methods lead to this: the party organization at first substitutes itself for the party as a whole; then the Central Committee substitutes itself for the organization; and finally a single 'dictator' substitutes himself for the Central Committee.
>
> (quoted in Deutscher, 1954, p. 90)

After Stalin's death power became less concentrated in the USSR. However, before *glasnost* there was an enormous concentration and fusion of economic and political power at the top of the state apparatus. There was a ruling group who sat on the Politburo and the various central committees and headed the various ministries and bureaus. This group, which comprised 'party officials, industrial administrators, diplomats, journalists, generals, secret policemen, trade union officials, artists and the occasional worker or peasant' (McAuley, 1977, p. 308), made decisions behind doors which were firmly closed to full-scale public scrutiny or widespread public discussion. Its members formed a privileged sector of Soviet

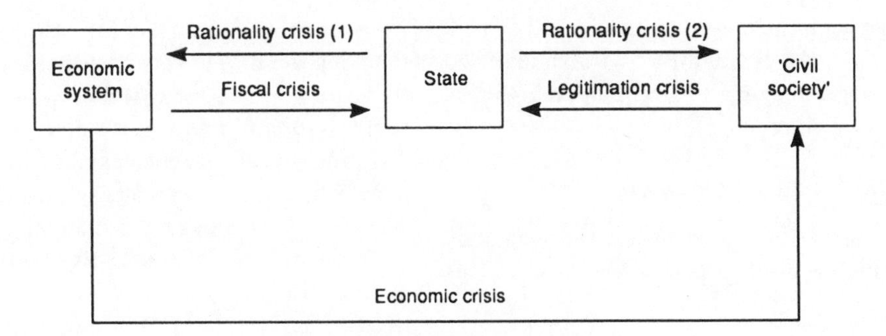

Figure 4.4 The crises of the state
Source: Short, 1984

society with more power and influence, more possessions and a more comfortable life-style than the mass of the population. The main problem in such a system was the inability of the economic system to meet the expressed needs of all the various social groups in different parts of the country. This is not a problem unique to the former USSR or socialist countries, but the heavy strain of arms expenditure and inefficient agriculture on the one hand and the variety of ethnic, religious and cultural groups on the other made the Soviet problems particularly acute.

Ideology

Revolutionary socialist parties have a vision of the future. The introduction of socialism, according to the theory, will usher in a new phase in the history of mankind. The advance of socialism will ultimately lead to communism, in which commodity production will cease, alienation will be overcome and new modes of human social behaviour will emerge. The advent of socialism, it is believed, will inaugurate a new social order and the transformation of the human condition. The state in socialist countries aims to aid this transformation by providing the conditions for the creation of a socialist ideology and a socialist people. The state becomes involved in art, religion, literature and all those areas of endeavour and practice which give substance to the present condition and intimations of future prospects. Culture cannot be separated from politics in socialist societies because culture is politics and politics is used to introduce a new culture.

At its best, culture in socialist societies can and did take on a new vitality, a new purpose. Bolshevik Russia before Stalin's rule provided examples of exciting developments in art, literature and music. At its worst, however, culture in socialist societies becomes a deformed agent of the state. The state becomes involved in cultural spheres for which it has little training and even less sensitivity. The deformation of art and the cinema in Russia under Stalin shows what happens when political conformity becomes the critical reference point. In closed socialist societies artistic and religious movements which fail to conform are perceived as a threat. The state is thus forced by its own momentum to react to every cultural and religious movement which does not toe the party line.

THE STATE AND CRISES

States throughout the world are involved in maintaining and reproducing the existing political and economic order. Apart from those brief revolutionary ruptures, when a new state seeks to destroy the basis of the old order, the primary goal of the state is to maintain the status quo. This is a difficult task. The world is constantly changing. The economic and political world order is constantly in flux. There is a need then to examine the major changes and subsequent responses through a discussion of crises. Figure 4.4 presents a simple model of crises between the economic system and civil society with the state as the mediating point.

In Chinese the word for crisis contains two words – danger and opportunity. This ambiguity is worth remembering.

Economic crises

An economic crisis occurs when the economic system fails to meet popular expectation. This occurs when:

- the world economy is in recession and unemployment rises
- the economic position of the individual country deteriorates

In capitalist countries an economic crisis is one of declining profits for capital and rising unemployment for labour. It is found in all countries to a certain extent when the world economy is in recession but is more pronounced in weaker economies in the core and it is endemic in the capitalist periphery.

The tension is more acute in the liberal democracies because the state is the arena of conflict between the competing forces of capital and labour. The state is charged to ensure the continued accumulation of profit but must also maintain popular support. In the short to medium term the state can pursue unpopular policies. Take the case of Britain. From 1945 to the 1970s it was losing its competitive position in the world economy. British companies were squeezed on the one hand by foreign competitors and on the other by the wage demands of organized labour. Glyn and Sutcliffe (1972) referred to it as *the profit squeeze.* The position worsened with the world recession in the 1980s . The 'solution' to the crisis emerged with the Thatcher government which came to power in 1979 with a policy of helping business and curtailing the power of the unions. In the early 1980s unemployment reached almost 3 million. The state had broken the back of organized labour and profitability was restored. In the long term, however, such a policy has its limits. Civil society becomes difficult to control with high levels of unemployment.

In capitalist states which are not liberal democracies unpopular strategies can be maintained over the long term. However, even in dictatorships there are limits to the extent to which a state can suppress its citizenry. Crises emerge elsewhere.

In the socialist state economic crises have been hidden by massive amounts of government intervention. Full employment is maintained at the cost of efficiency. The economic crisis becomes a legitimation crisis.

Legitimation crisis

A legitimation crisis occurs when the state cannot maintain the necessary degree of popular loyalty. The state loses credibility and ultimately loses the ability to govern.

States experience legitimation crises when they pursue unpopular policies for too long. There are limits to which a state can reflect only the needs of a small minority, whether this minority be the cronies of corrupt dictators like August Somoza and Ferdinand Marcos or members of the ruling communist party in a one-party state. States which experience legitimation crises need a large military and police presence. This can come from outside; in the past the USA was always willing to help right-wing dictators facing a popular uprising if the conflict could be dressed up as a fight against communism, while the Soviets could be relied upon to send in the tanks if any communist party looked as if it was about to be toppled. However, the brutal facts of imperial overstretch led to a Soviet retreat from empire.

A legitimation crisis can be solved in the short to medium term by naked repression. The state has immense powers. As Tolstoy once said, 'No one who has not sat in prison knows what the state is like'. Malevolent states exist in the world and the annual reports from Amnesty International reveal innumerable cases of oppression, imprisonment and execution. State terror is used to prop up unpopular regimes and silence the voice of critics.

In the long term, however, states which lack popular support find it difficult to remain in power. An alienated population can become the source of popular protest (see chapters 7 and 8).

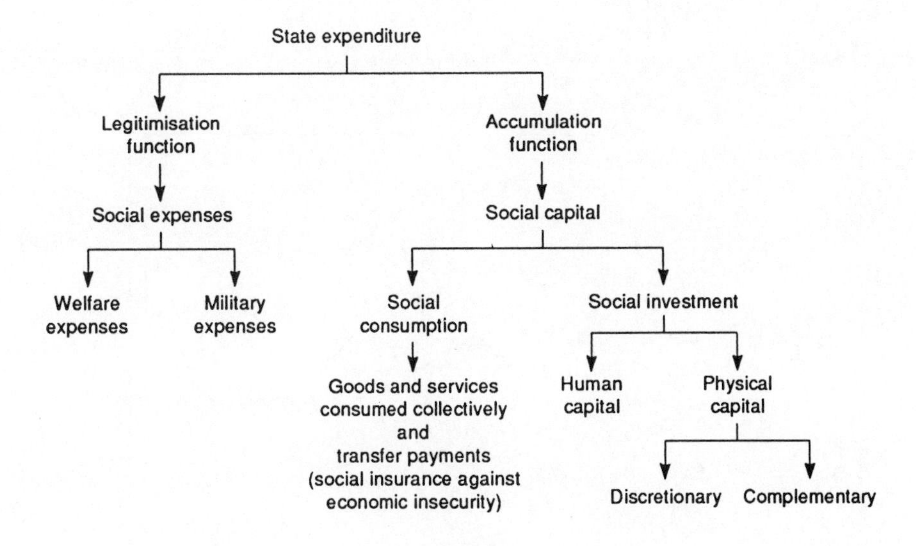

Figure 4.5 Categorization of state expenditures

Fiscal crisis

A fiscal crisis of the state occurs when expenditures exceed revenue. The fiscal of the capitalist state was the subject of analysis by James O'Connor (1973). He asserts that the state must fulfil two basic, but contradictory, functions. It must ensure the conditions of capital accumulation and maintain the basis for its own legitimation. Each of these two functions is associated with differing types of public expenditure (see Figure 4.5).

Social capital is that expenditure required for capital accumulation; it takes the form of social investment and social consumption:

- Social investment expenditures involve those projects and services which increase the productivity of a given amount of labour power; they include capital investment in physical infrastructure (roads, railways, airports, etc.) and investment in research and development and education. Social investment is increasing because the growing size and complexity of production involves large and costly infrastructure investment.
- Social consumption expenditures include investment in projects and services which lower the costs of reproducing labour power. Two types are identified by O'Connor: those public sector goods which are collectively consumed by working people (e.g. national health services) and social insurance expenditures for the short-term unemployed. Social consumption expenditures have been growing in response to organized demands for better welfare and as a function of the state's need to ensure mass loyalty.

Social expenses expenditures consist of spending on projects and services which are required to maintain social harmony. Two types are identified in the USA:

- military expenses, which grew in the post-war period because of the growth of the military-industrial complex
- welfare expenditures, which are designed for the long-term unemployed and other elements of the population 'surplus' to economic requirements. This latter form of expenditure has been growing because of the introduction of labour-saving devices. Superimposed upon this general trend there are secondary peaks and troughs as welfare expenditure expands

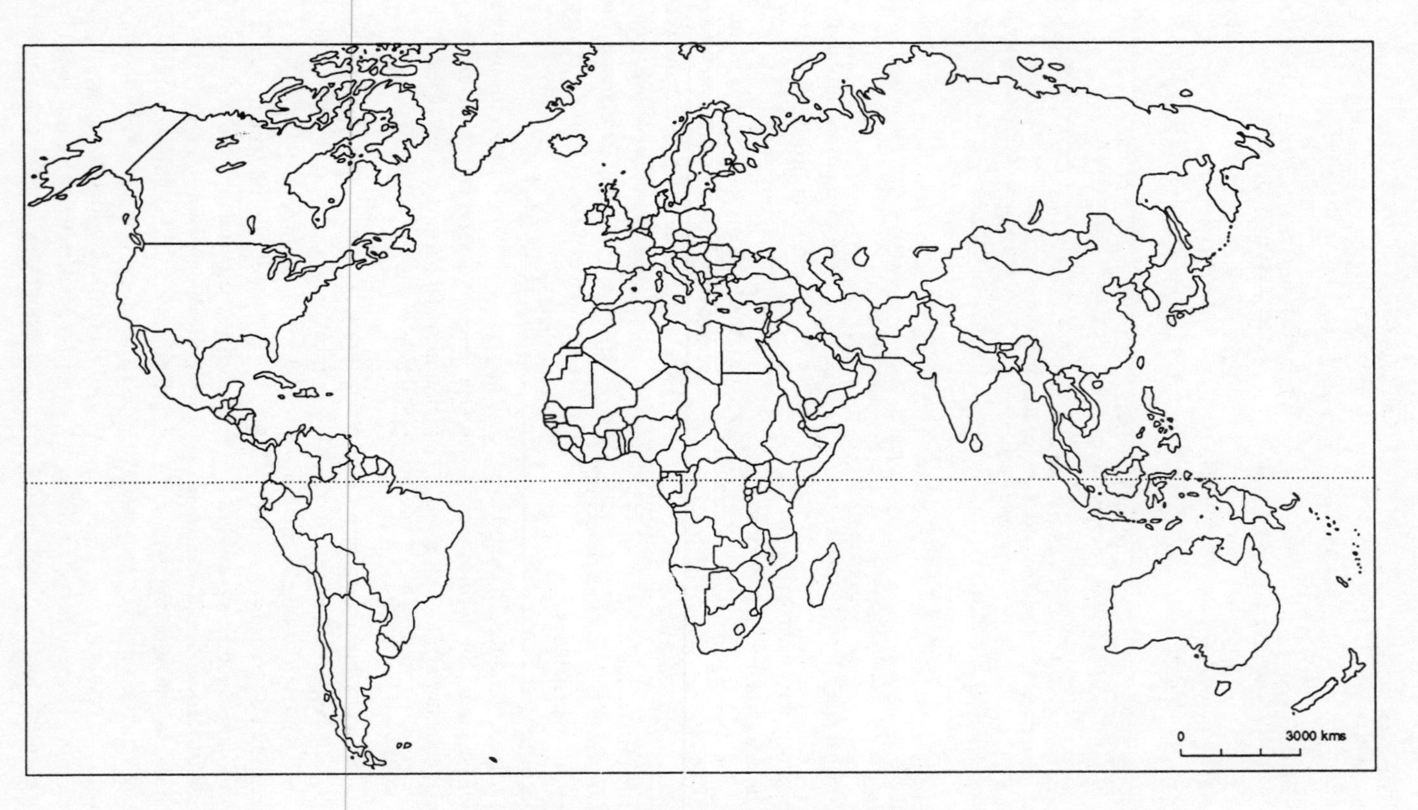

Figure K.1 Selected countries of the world, 1990

BOX K:

Test your knowledge of the countries of the world. Figure K.1 is a outline map, name the countries then check your answers in an atlas.

Alternatively, use Table K1 and see how many of the countries cited you can locate on the outline map. Just to keep you on your toes there are several deliberate errors in Table K1. Can you identify them?

Tally up your score. Give yourself one point for each correct answer.

Over 100	–	You sure you didn't cheat?
Over 50	–	You should be writing this book.
25–50	–	Pretty good.
10–25	–	OK, but could do better.
5–10	–	Are you sure you are reading the right book? Are you doing another course? Economics maybe?
Less than 5	–	Oh dear, oh dear, oh dear.

Table K1

1. Afghanistan
2. Albania
3. Algeria
4. Angola
5. Argentina
6. Armenia
7. Austria
8. Azerbaijan
9. Bangladesh
10. Belarus
11. Belgium
12. Belize
13. Bolivia
14. Borneo
15. Bosnia and Hercegovena
16. Botswana
17. Brazil
18. Bulgaria
19. Burma
20. Burundi
21. Cambodia
22. Cameroon
23. Canada
24. Central African Rep.
25. Chad
26. Chile
27. China
28. Colombia
29. Commonwealth of Independent States
30. Congo
31. Costa Rica
32. Croatia
33. Cuba
34. Cyprus
35. Czechoslovakia
36. Dahomey
37. Denmark
38. Dominican Rep.
39. Ecuador
40. Egypt
41. Eire
42. El Salvador
43. Equatorial Guinea
44. Estonia
45. Ethiopia
46. Falkland Islands
47. Finland
48. France
49. French Guiana
50. Gabon
51. Gambia
52. Germany
53. Georgia
54. Ghana
55. Greece
56. Greenland
57. Guatemala
58. Guinea
59. Guyana
60. Haiti
61. Honduras
62. Hong Kong
63. Hungary
64. Iceland
65. India
66. Indonesia
67. Iran
68. Iraq
69. Israel
70. Italy
71. Ivory Coast
72. Jamaica
73. Japan
74. Jordan
75. Kazakhstan
76. Kenya
77. Kirgiz
78. Laos
79. Latvia
80. Lebanon
81. Lesotho
82. Liberia
83. Libya
84. Luxembourg
85. Madagascar
86. Malawi
87. Malaysia
88. Mali
89. Malta
90. Mauritania
91. Mauritius
92. Mexico
93. Moldavia
94. Mongolia
95. Morocco
96. Mozambique
97. Nepal
98. Netherlands
99. New Zealand
100. Nicaragua
101. Niger
102. Nigeria
103. North Korea
104. Norway
105. Oman
106. Pakistan
107. Panama
108. Paraguay
109. Peru
110. Philippines
111. Poland
112. Port Guinea
113. Portugal
114. Romania
115. Russia
116. Rwanda
117. Saudi Arabia
118. Senegal
119. Seychelles
120. Sierra Leone
121. Singapore
122. Slovenia
123. Somali Rep.
124. South Africa
125. South Korea
126. South Yemen
127. Spain
128. Sudan
129. Surinam
130. Swaziland
131. Sweden
132. Switzerland
133. Syria
134. Tadzhikistan
135. Tanzania
136. Thailand
137. Togo
138. Tunisia
139. Turkey
140. Turkmenistan
141. Uganda
142. Ukraine
143. United Arab Emirates
144. United Kingdom
145. Uruguay
146. USA
147. Uzbekistan
148. Venezuela
149. Vietnam
150. West Sahara
151. Yemen
152. Yugoslavia
153. Zaire
154. Zambia
155. Zimbabwe

during periods of social unrest and civil disturbance and contracts when political and social stability is restored (Piven and Cloward, 1977).

According to O'Connor the structural gap between state expenditure and revenue is caused by the socialization of increasing costs and the private appropriation of profits. It takes the form of a crisis because the gap between expenditure and revenue can only be closed at the expense of major redistributional consequences. On one side of the fiscal equation, reduction in expenditure becomes difficult because business wants social investment to be maintained if profits are to be assured, but a major reduction in welfare expenditure will raise the problem of legitimation for the state. On the other side, increases in state revenue are difficult to maintain because business interests lobby against company taxation eating into profits and high levels of personal taxation are difficult to maintain in a democracy.

The crisis has become apparent in a number of ways in Western Europe and North America:

- state expenditure cutbacks have been directed towards welfare expenditures, affecting those groups with least political power and little economic muscle, and social consumption expenditures, where the relationship with capital accumulation is more long-term and less immediately apparent
- there have been attempts to reduce the number of public-service workers and to streamline working arrangements in the public sector
- the crisis has been manifested in popular demands for reduction of personal taxation
- the crisis has appeared differently at various levels of the state. In the USA the central government appropriates the majority of tax revenues but the state and local governments have to provide welfare expenses and elements of social capital. The mismatch between revenue and expenditure is thus severe at the local government level and most acute at the city level, where welfare expenditures are greatest. The fiscal crisis of major US cities is the spatial manifestation of the fiscal crisis of the state.

Rationality crises

Two forms of rationality crisis can be identified. Type 1 occurs when the government gets its economic policies wrong. In order to avoid economic crises, states become involved in the economic order. Even in the capitalist world the state is involved in a vast number of interventions, from providing legal frameworks, to setting the exchange rate. This type of rationality crisis is the failure to produce the requisite number of correct decisions. It is particularly acute in centrally planned economies because here the state makes most of the decisions. The state is called upon to make investment decisions, produce and sell goods. Errors became exaggerated through the system and can lead to bottlenecks, and, particularly important for maintaining popular support, a failure to produce the expected quality and quantity of consumer goods. If people have to stand in line for too long to buy essential goods and services then a rationality crisis can produce a legitimation crisis.

Type 2 occurs in the variety of social welfare policies and is apparent in such things as the failure of the public education system to produce people with the requisite skills for the job market.

The two types are closely related as economic policies have implications for social welfare which in turn influence the economic system. If the public education system is underfunded because of failure of the state to invest enough money (Type 1) then the educational sector may fail to produce enough people with the right skills for the job market (Type 2) which leads to problems of a skilled labour shortage and hence the possibility of an economic crisis.

Motivation crises

A motivation crisis has been defined by Jurgen Habermas as the 'discrepancy between the need for motives declared by the state, the educational system and the occupational system on the one hand, and the motivation supplied by the socio-cultural system on the other' (Habermas, 1976, p. 75). Habermas asserts that advanced capitalist societies now produce social groups and ideas which are set at variance with the ideological basis of capitalism. The radicalism of the professions, the growth of a large sector of the population which is economically inactive (e.g. students, unemployed and retired) and associated changes in attitudes towards individual achievement, work and leisure are producing a whole belief system which challenges the old justifications for capitalism. Ideas are developing in advanced capitalist societies, Habermas seems to be arguing, which have the potential ability to disrupt and even threaten the continued existence of capitalism. The economic system of the advanced capitalist countries is based on the work ethic and the continued consumption of goods and services. If either of these two ideologies are displaced from their central positions then the system could be in danger of collapse. In terms of the work ethic, people may not want to work but they have very little alternative if they are to survive. The ideology of conspicuous consumption however, seems to be less secure. In the last decade of the twentieth century more and more people are questioning the need for wasteful consumption. A green revolution, involving less rather than more consumption, may be just around the corner.

Throughout the 1980s and early 1990s the main motivation crisis was faced by the states of the socialist bloc. Lacking incentives or penalties the economic system failed to generate the enthusiasm or the full activity of the population. *Perestroika*, or restructuring, was a Soviet attempt at improving motivation. In China market forces have been allowed to develop and grow since the late 1970s. Throughout Eastern Europe experiments in freeing the economy from the monopoly grip of a central state are now well underway. There experiments are not without problems. Populations used to full employment and low prices may want the benefits of market forces (more consumer durables, greater purchasing choice) but they do not want the disadvantages (unemployment, glaring inequalities). The trick is to discover how you get the one without the other.

The world around us, the world of physical aspects like houses, tables and chairs looks solid and immutable. At another level, the sub-atomic level, particles are constantly moving and vibrating. So it is with the world order. Behind the world *order* (even the name implies some form of coherence and stability) states are constantly adjusting. The changes include the long cycles of superpower growth and decline and the shorter cycles of responses to crises. The world order of states should more accurately be described as the world flux of states.

GUIDE TO FURTHER READING.

Descriptions of superpower growth and decline include:

Kennedy, P. (1987) *The Rise and Fall of The Great Powers: Economic Change and Military Conflict from 1500 to 2000.* Random House, New York.

Modelski, G. (1983) 'Long Cycles of World Leadership', in W. R. Thompson (ed) *Contending Approaches to World System Analysis.* Sage, Beverly Hills.

Modelski, G. (1987) *Long Cycles in World Politics.* Macmillan, London.

On different approaches to the state in advanced capitalist countries see:

Dahl, R. A. (1982) *Dilemmas of Pluralist Democracy.* Yale University Press, New Haven.

Dahl, R. A. (1989) *Democracy and its Critics.* Yale University Press, New Haven.

de Jassay, A. (1985) *The State.* Basil Blackwell, Oxford.

Galbraith, K. (1967) *The New Industrial State.* Hamish Hamilton, London.

Jessop, B. (1982) *The Capitalist State.* Martin Robertson, Oxford.

Kiny, R. (ed) (1983) *Capital and Politics.* Routledge & Kegan Paul, London.

Miliband, R. (1969) *The State in Capitalist Society.* Weidenfeld & Nicolson, London.

Miliband, R. (1983) *Class Power and State Power.* Verso, London.

Piven, F. and Cloward, R. (1982) *The Class War.* Pantheon, New York.

Self, P. (1985) *Political Theories of Modern Governments.* Allen & Unwin, London.

On the state in the periphery have a look at:

Alavi, H. (1972) 'The state in post colonial society', *New Left Review*, 74, 59–81.
Alavi, H. and Harriss, J. (eds) (1989) *Sociology of 'Developing Societies': South Asia.* Macmillan, London.
Cammack, P., Pool, D. and Tordoff, W. (1988) *Third World Politics.* Macmillan, London.
Goulbourn, H. (eds) (1980) *Politics and The State in The Third World.* Macmillan, London.
Munck, R. (1979) 'State and capital in dependent social formations', *Capital and Class*, 8, 34–53.
Munck, R. (1984) *Politics and Dependency in The Third World.* Zed Press, London.
Randall, V. and Theobald, R. (1985) *Political Change and Underdevelopment.* Macmillan, London.
Tordoff, W. (1984) *Government and Politics in Africa.* Macmillan, London.
Wynia, G. W. (1984) *The Politics of Latin American Development.* Cambridge University Press, Cambridge.

On the state in the socialist world see:

McAuley, M. (1977) *Politics and The Soviet Union.* Penguin, Harmondsworth.
Bahro, R. (1978) *The Alternative in Eastern Europe.* New Left Books, London.
Harding, N. (ed) (1984) *The State in Socialist Society.* Macmillan, London.
Sakwa, R. (1989) *Soviet Politics: An Introduction.* Routledge, London.
Furtak, R. K. (1986) *The Political Systems of The Socialist States.* Wheatsheaf, Brighton.

A useful introductory reader is:

Held, D. *et al.* (eds) (1983) *States and Societies.* Martin Robertson, Oxford.

For a critical account of Marxist theories see:

Dunleavy, P. (1985) 'Political Theory', in Z. Baranski and J. R. Short, (eds) *Developing Contemporary Marxism.* Macmillan, London.

On crisis see:

Habermas, J. (1976) *Legitimation Crisis.* Heinemann, London.
O'Connor, J. (1973) *The Fiscal Crisis of The State.* St Martin's Press, New York.
O'Connor, J. (1981) 'The fiscal crisis of the state revisited', *Kapital state*, 9, 41–62.
Offe, C. (1984) *Contradiction of The Welfare State.* Hutchinson, London.
Short, J. R. (1984) *The Urban Arena.* Macmillan, London.

Relevant journals

American Journal of Political Science
American Political Science Review
British Journal of Political Science
Journal of Politics
Journal of Public Policy
Political Studies

Other works cited in this chapter

Bachrach, P. and Baratz, M. S. (1963) 'Decisions and non-decisions: an analytical framework', *American Political Science Review* 57, 641–51.
Baran, P. and Sweezy, P. M. (1960) *Monopoly Capital.* Monthly Review Press, New York.
Dahl, R. A. (1956) *A Preface to Democratic Theory.* University of Chicago Press, Chicago.
Deutscher, I. (1954) *The Prophet Armed.* Oxford University Press, Oxford.
Glyn, A. and Sutcliffe, B. (1972) *British Capitalism, Workers and The Project Squeeze.* Penguin, Harmondsworth.
Nove, A. (1980) 'Problems and prospects of the Soviet economy', *New Left Review*, 119, 3–19.
Lenin, V. I. (1902; 1965) What Is To Be Done. Progress Publishers, Moscow.
Piven, F. F. and Cloward, R. (1977) *Poor People's Movements: Why They Succeed, How They Fail.* Pantheon, New York.
Trotsky, L. (1931; 1977) *The History of The Russian Revolution.* Pluto Press, London.

5

THE NATION-STATE

> Essentially, nationalism is a phenomenon connected not so much with industrialization or modernization as such, but with its uneven development.
>
> (Gellner, 1964)

NATIONS AND STATES

A distinction can be made between nations and states: a *nation* is a community of people with a common identity, shared cultural values and an attachment to a particular territory. Their identity is intimately associated with the territory. A *state*, on the other hand, is a political organization covering a particular territory.

The state 'is likely to show the greatest stability and permanence when it corresponds closely with a nation' (Pounds, 1972, p. 12). The correspondence, let us call it congruence, depends upon a number of factors. If the state boundaries enclose a single language group, sharing the same religion and cultural heritage, then, holding everything else constant, the stability of the nation-state is assured. If, on the other hand, the state encloses a variety of different language groups, diverse religious sects and sections of the population with dissimilar cultural traditions, all in separate and distinct areas, then political stability is not so easily maintained. Congruence occurs when there is a perfect fit between nation and state. It is a rare occurrence. Incongruence between nation and state is more common. Two different types of incongruence can be noted:

Nations without states

Throughout the world there are instances of nations without states; they are the dispossessed, people whose shared values and aspirations are not expressed in state formation. Examples include the Palestinians and the Kurds. Figure 5.1 shows the distribution of the Kurds, a people whose 'national' territory has been divided up by different state boundaries. The result is a nation without a state. The Kurds, like other such groups, are the losers in the game of international politics. They live in the states that are controlled by other people. Their tragedy is that they have few oppor-

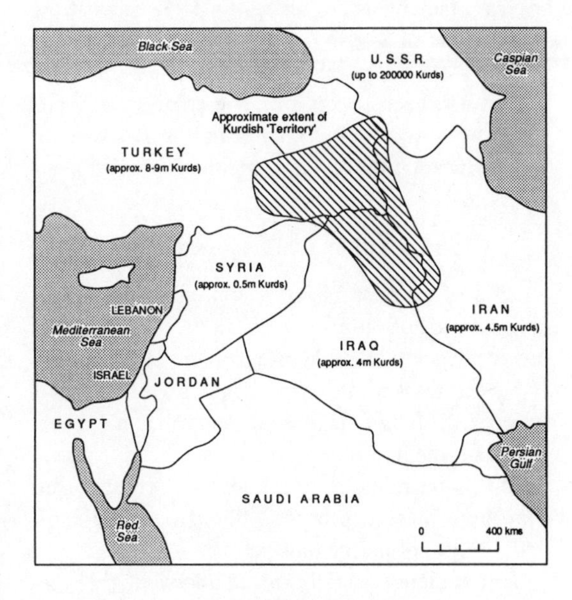

Figure 5.1 The Kurdish 'nation'

tunities to present their case. The forum of world opinion is more responsive to states than to dispossessed nations. Nations without states lack the formal channels of representation. This is one reason behind the tactics of terror sometimes adopted by such dispossessed groups. Indeed, the definition of terrorism is often used to describe the actions of these groups. When a state bombs another state it is war, when unrepresented national groups do the same thing it is called terrorism.

Some nations do achieve their dream. The most recent example has one of the longest histories. Since the destruction of Jerusalem by Nebuchadnezzar in 586 BC, the Jews have been dispersed throughout the ancient world and subsequently throughout all the world. Yet they retained their religion, their history and a vision of a Jewish homeland. This belief was strengthened throughout the nineteenth and twentieth centuries as more Jews moved to Palestine. The belief was given extra impetus by the unimaginable horror of the Holocaust, in which the Nazis sought to eradicate a whole race of people. They almost succeeded: estimates vary, but almost 6 million Jews were murdered. The desire to establish a presence in the ancient homeland grew in the post-war era. Israel was established in 1947 against the background of British control and Arab resistance. A new state was established (see chapter 6). One nation had been successful. The creation of Israel gave the Jews a state but meant the Palestinians were now a nation within someone else's state.

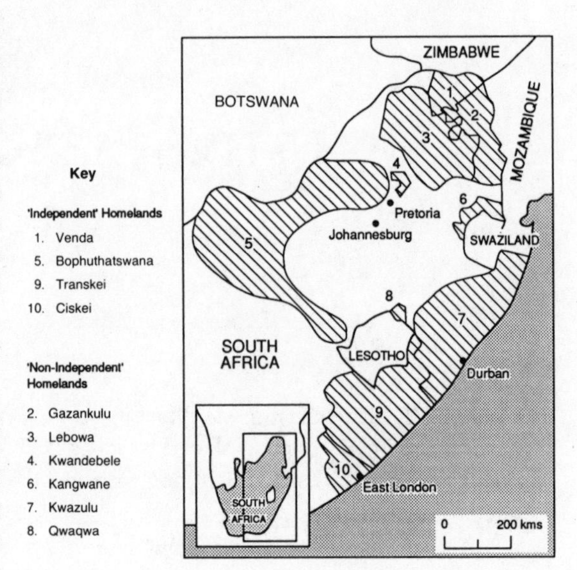

Figure 5.2 Black homelands in South Africa, 1990

States with more than one nation

Israel was an example of a common occurrence, one state but a number of nations. This form of incongruence is the rule rather than the exception. It is less marked in some countries of the new world, e.g. Brazil, USA and Australia, no doubt reflecting the initial destruction or marginalization of the indigenous people and their culture. But even here there has been a 'rebirth' of the territorial claims of indigenous people.

Incongruence is most pronounced in the very old states. In Europe state boundaries evolved as certain regional centres of power grew and expanded while others declined. State boundaries overlay the tight mosaic of differing ethnic formation. The result was a multi-ethnic state. In the case of the United Kingdom, the state incorporated the Welsh, Irish and Scots as well as the English. Across the Channel, Normandy, Brittany, Provence, Burgundy and the Midi were incorporated into the French state. In the case of Russia, tsarist expansionism eastward and southward meant that the Russian state, and eventually the Soviet state, contained a varied ethnic mix including Georgians, Armenians, Azerbaijanis, Turkomen and Mongols.

Incongruence also occurs in the very newest states. In the periphery of the world economy states grew from a colonial basis of territorial demarcation rather than one of ethnic cohesion. The subsequent history of sub-Saharan Africa, for example, is one of tribal tension sometimes erupting into open conflict. One example is the attempt by the Ibos to create the separate state of Biafra from Nigeria in 1967. Thirty months of bloody civil war ended in 1970.

State boundaries often incorporate a variety of

nations. The state can adopt various strategies to deal with this situation:

Repression

After the battle of Culloden in 1746 the Highlands of Scotland came under the control of the Hanoverian state, centred in London. The basis of Highland identity was repressed; it became illegal to wear the plaid, play the bagpipes or speak in Gaelic. The state sought to repress the basis of Gaelic national identity.

Ever since 1778 the original inhabitants of Australia, the Aborigines, have been badly treated by the white (predominantly British) settlers. No treaty was ever signed, the land was stolen and the Aborigines were denied effective rights. They were not counted in the census until 1967 or allowed to vote. Their national identity was almost destroyed, denied and marginalized. In the past 10–15 years, however, aboriginal resistance has been reactivated. Ayer's Rock has been handed back and substantial land claims have been successful. More generally, there has been an acceptance of the rights of the indigenous people.

Indirect repression

This can occur through the steady erosion of cultural identity by the use of a common language and creation of a national education system. Repression can also occur through the tight confinement of ethnic groups to specific parts of the country. The apartheid system of South Africa was based on racist legislation and a brutal use of state power. Blacks were confined to specific territories and confined to particular parts of cities (see Figure 5.2).

Accommodation

In the nineteenth and throughout most of the twentieth century most states sought to centralize and modernize. In Europe ethnic differences were discouraged. In Britain, for example, Gaelic and Welsh were discouraged. At the end of the twentieth century, however there has been an acceptance, willing or unwilling, of the need to accommodate these differences while still retaining territorial integrity. In many countries of the world there is now a greater accommodation of ethnic and linguistic differences. In Barcelona, Spain, the local trains have 'no smoking' signs in both Spanish (the language of Madrid and Castille) and Catalan (the language of Barcelona and the surrounding district). When the British government established a new television channel (Channel 4) Welsh representation was successful in getting a Welsh-speaking component. In south-east France one can see town signs in Provençal as well as French.

Accommodation can be a dangerous strategy. It may buy off ethnic discontent but it may also provide the basis for social movements which question the territorial integrity of the existing state.

Federalism

The most complete form of accommodation is to have separate government institutions for different parts of the territory of the state. The federal solution divides up the power and authority of the state to separate regional authorities. It is a very rare strategy as most central authorities are loath to devolve power, while most national movements want independence not federation.

Towards congruence?

The state is a powerful agent. It taxes the population, runs the army, controls law and order and the education system and has some measure of control over the mass media. It seeks to control the information and social messages passing through a society. This immense power is used to inculcate in its citizens a belief in the state. When school children throughout the world stand to attention for their national anthem or salute their national flag they are reproducing the belief in the state.

States use their power to weld citizens into a coherent society with similar values and belief systems. Some are more successful than others and some more sophisticated than others but the general goal is the same. In the long term, therefore, we would expect to find increasing congruence between the nations within a single state. However, there are forces which militate against this. These include:

(a) *Language*: if different languages are spoken within one state congruence can only be achieved if a common language emerges. In Britain, English has emerged and the use of Welsh and Gaelic has declined. Elsewhere, however, language differences persist. In Spain, for example, Catalan and Basque are the everyday languages of people in specific regions, kept alive by tradition and custom, fanned by writers and intellectuals.
(b) *Religion*: Separate religion may not cause separate movements. The rich variety of religious observance in the USA is not the basis of social conflict. However, it may exacerbate differences, as in Northern Ireland between the pro-British Protestants and the Republican Catholics and in Lebanon where the Muslim/Christian distinction is just one more line of fracture between warring communities.
(c) *Economic differences*: If long term economic differences overlie ethnic differences then congruence may be very difficult to achieve, especially if the minority group perceives economic inequality. In Britain, for example, Scottish and Welsh nationalism has been kept alive more by economic inequalities than by cultural resistance. Constantly higher unemployment levels and lower average incomes in the two regions compared to the south-east of England have given a backbone of discontent to nationalist sentiments. Some indication of the nationalist feeling in Europe is given in Figure 5.3. Let us look at this relationship between uneven economic development and nationalism in more detail.

Uneven development and nationalism

Uneven development predates capitalism but the links between uneven development, which we can define as the geographical unevenness of economic growth and development, and nationalism were created by capitalism. The industrial revolution began in England around 1780. Thereafter, as the effects of the transformation spiralled outward from the storm centre the growth of nationalism arose in the wake of the storm; more than this, it was part of the storm. We can identify two phases. The first is associated with the rise of nationalism in nineteenth- and early twentieth-century Europe: first Germany, then Italy and then the Balkans and Eastern Europe. Here the growth of capitalism was associated with the slow transformation of agriculture and the steady growth in industrialization. The geographical unevenness of the process was shaped by the nature of the resource endowment, the character of the pre-capitalist social formations and the differential speed of the diffusion of capitalist relations of production. Subsequently, the spatial differentiation produced by these processes provided the context for later rounds of capital investment.

Uneven development in Europe took place against a mosaic of different linguistic, ethnic and cultural regions best exemplified in the diversity of the constituent elements of such ramshackle outfits as the Austro-Hungarian and Ottoman empires. It is against such a background that perceived regional inequalities began to develop into nationalism. Gellner (1964) presents a simple model of the process. If two large regions are differentially affected by capitalist development, one becoming richer than the other, but the two regions do not differ in culture or language, then it is unlikely that regional discontent will be manifest in separatist movements. If the regions do differ, then the superimposition of cultural and economic differences provides the necessary preconditions for the growth of nationalist sentiment. For the sentiment to foster political movements with pertinent effects it is important to have a discontented intelligentsia to give cultural focus, differences in language, a

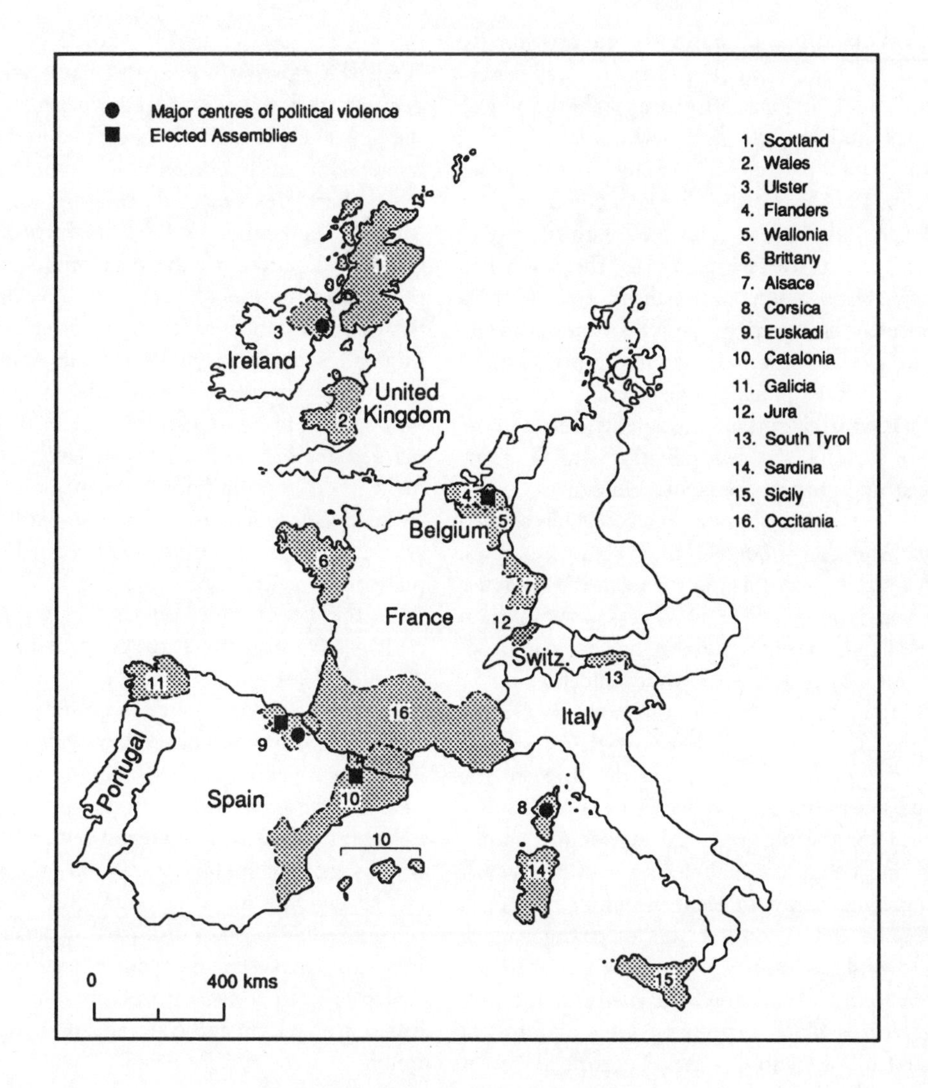

Figure 5.3 Minority areas in Europe

broad base of workers and peasants to give the weight of numbers to protests, and a stupid central authority which leaves little room for nationalist sentiment to develop within existing political structures.

In one sense nationalism can be seen as the cultural fall-out from the economic explosion of capitalist uneven development. In another, nationalism was not only a reaction to perceived regional inequality but an attempt to guide economic development; it was the ideological rationale for the drive from semi-periphery to core. In southern, central and eastern Europe nationalism was the language used to justify the attempts to shift from semi-periphery to core status in the world economy. Nationalism and economic motives were fused into a form of economic nationalism which gave cultural substance to economic drives and economic goals for cultural forms. Nationalism was not only a reaction to uneven development, it was the cultural manifestation of the attempts to guide this development. The ideas are well articulated in the German Historical School, whose members outlined the neo-mercantilist creed of economic nationalism for a rising Germany (see Lichteim, 1974, especially Chapters 5 and 6).

The second phase is associated with the rise of nationalism in the periphery in the twentieth century. Here again nationalism was both a response to the nature of capitalist uneven development and an attempt to guide the processes of economic development. Liberation movements were national liberation movements because the fight for national self-determination was the same as the struggle for economic independence.

In an arresting metaphor Tom Nairn has termed nationalism the modern Janus (Nairn, 1977). Nairn's argument is this: nationalism is a response to uneven development. The process is twofold: on the one hand, the elites mobilize against the 'progress' which destroys the old order and they do so by calling up their own cultural resources. On the other hand, they try to improve their position within the world economy, and the drive to the semi-periphery or core status uses nationalism as a platform for the big push. All nationalist movements thus contain a dual character of backward-looking, almost atavistic elements which hark back to the past and dynamic forces of change which point toward the future.

We can think of a series of *centripetal* and *centrifugal* forces which respectively unify and disrupt the control of central state. Centripetal forces include such things as external aggression which unifies the population, sensitive federal structures which allow the culture and tradition of different parts of the country to be 'safely' expressed, an education system which socializes all children into a 'national', as opposed to a local or separatist, ideology, and mass media which successfully inculcates the population into a belief in the essential rightness of the present governmental arrangements.

The most powerful centrifugal force is perceived political and economic inequality. Secessionist movements tend to gain large-scale support when the inequalities are widely and deeply perceived, when the existing authorities have failed or are deemed to have failed to improve matters, when the central government is associated in the people's minds with a different cultural or ethnic group and when there are perceived economic opportunities (or few hardships) to be gained by separation. Centrifugal forces are given life and substance by the uneven nature of economic development. In these circumstances some of the things which were originally classified as centripetal can become centrifugal in effect. The mass media tend to homogenize tastes and expectations, but if there are marked regional differences in levels of achievement then the unifying influence of the mass media becomes a divisive force. Similarly, a federal system devised to promote unity can become the platform for secession.

In the richer core countries of the world the centripetal forces tend to be stronger than the centrifugal pressures. This is partly a function of time. The boundaries of the core states, with some notable exceptions, have very largely remained intact for over a hundred years. The passing of time has the same effects on international boundaries as it does on accumulated wealth; although the origins may be shabby and bloody the process of ageing gives an air of respectability. The strength of centripetal forces is also partly a function of the powerful mass media and education systems, which tend to round out the contours of cultural differences within the state and promote national unity.

Centrifugal forces tend to be strong in the periphery and semi-periphery. In the peripheral and semi-peripheral countries of Africa and Asia the present international boundaries tend to be younger than those in the core. In contrast to the first phase of nationalism, the second has a much weaker cultural base. Very often there was no national history of cultural cohesiveness in the areas which become dependent states; independence movements, especially those in Africa, tended to work within the colonial administrative boundaries. After independence had been achieved, the cohesion of the anti-colonial struggle

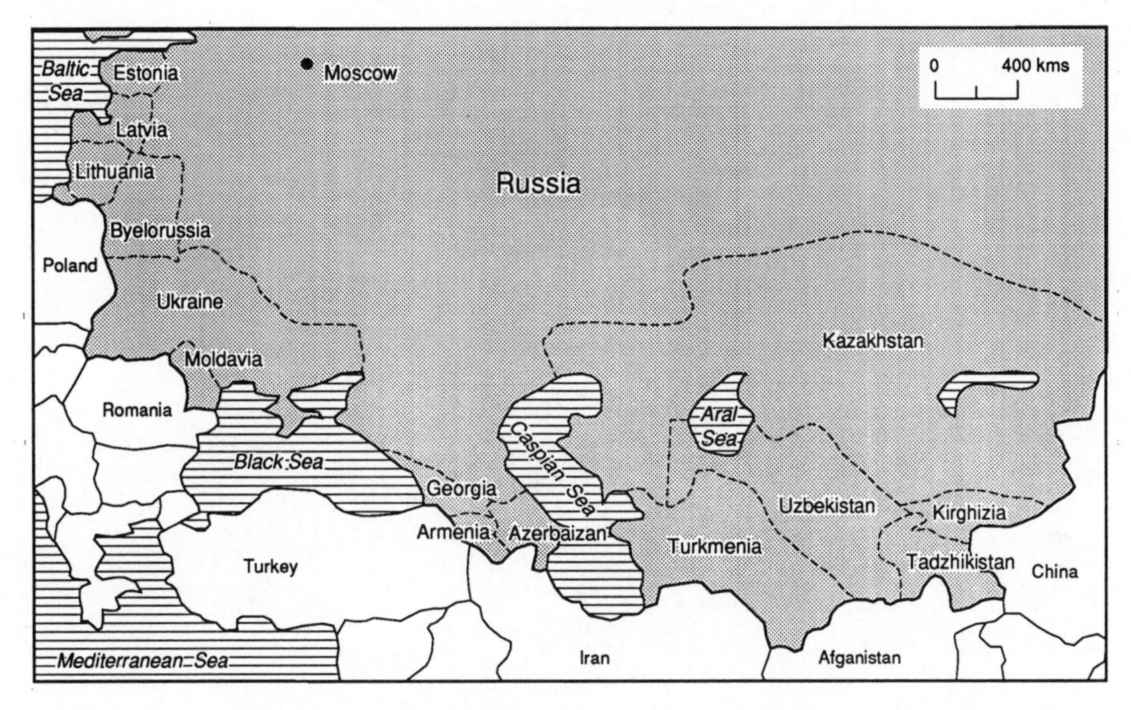

Figure 5.4 Republics of the Soviet Union, 1990

tended to fracture along older cultural fault-lines and politics in post-independence sub-Saharan Africa have been dominated by tribalism.

The centrifugal forces are given added impetus by the nature of incorporation into the capitalist economy. The importance of primary products to the economies means that basic economic development is associated with the uneven pattern of resource endowment. Moreover, the growth of the manufacturing and service sectors (especially marked in the semi-periphery) tends to be highly concentrated in the large cities of certain regions. There is little opportunity for regional policy because the state has limited power with regard to the multinational companies.

Although uneven development reaches its most pronounced expression in the periphery, regional differences are not always expressed in separatist movements. In those countries such as Tanzania with a large number of very small tribal groupings there is no cultural base large enough to mobilize an alternative to the present state. There is no necessary one-to-one correspondence between uneven development and nationalism.

In the socialist states economic centrifugal forces are supposed to be minimized by the use of central planning, which aims to eradicate regional inequality. However, central planning failed to overcome the legacy of regional differences in resource endowment, levels of urbanization and industrialization. Besides, the early emphasis of socialist planning on heavy industry tended to increase inequalities between regions and between towns and cities, and there are still marked regional differences within and between socialist states. Many former socialist states contain a variety of different nations. In Yugoslavia the incendiary mixture of Croats, Slovenes, Serbs and Magyars finally exploded into civil war in 1991. In the Soviet Union the diverse nationalities of Estonians, Lithuanians, Latvians, Georgians, Armenians and others, was a powerful centrifugal force in the final collapse of the Soviet Union in 1991 (see Figure 5.4).

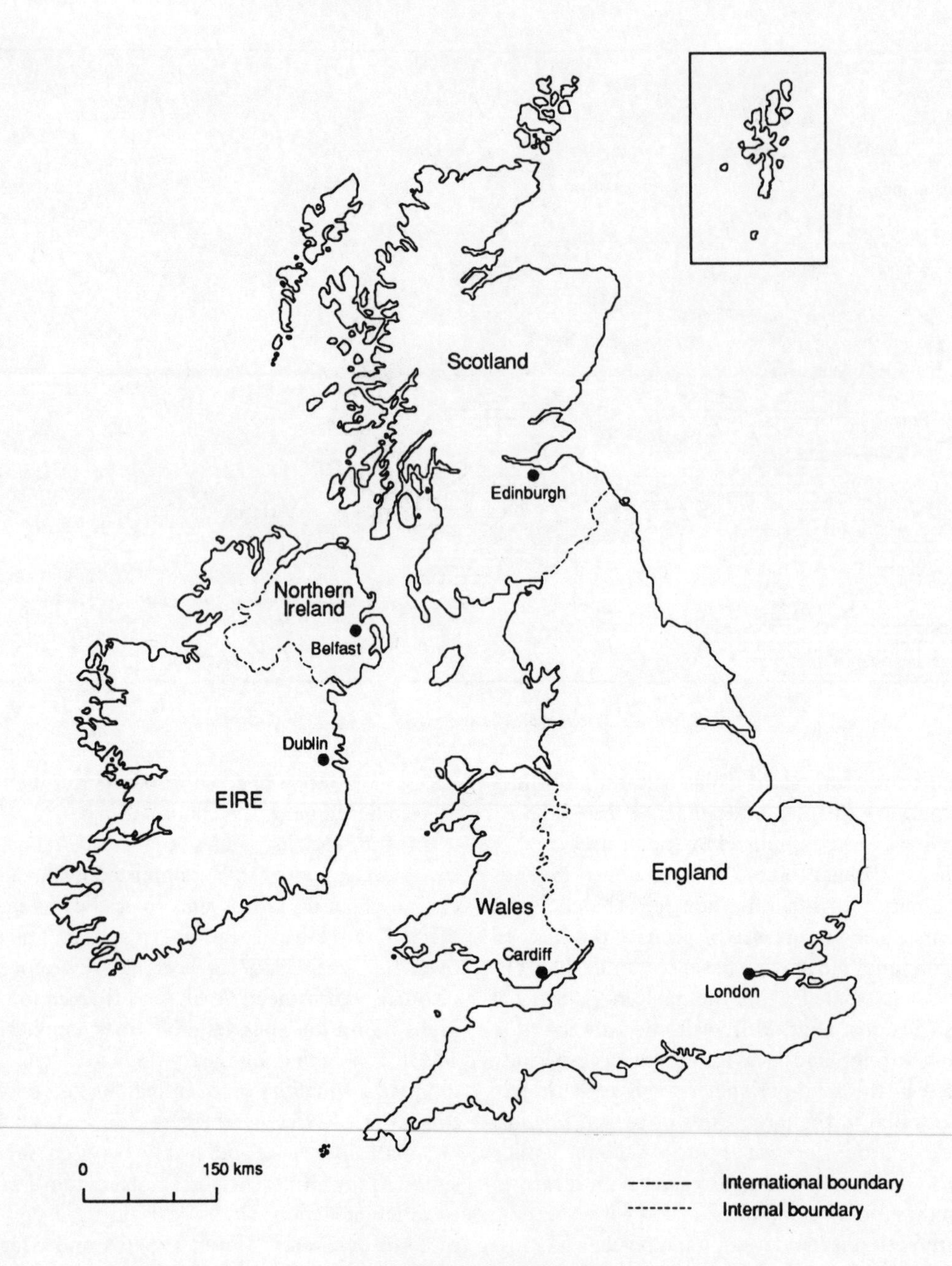

Figure 5.5 The United Kingdom

THE SPATIAL ORGANIZATION OF THE STATE

Each state has formal control over part of the surface of the earth. The form of this occupancy varies. We can identify two model types:

- the unitary state
- the federal state

Let us look at each in turn, bearing in mind that these are ideal types and that in reality states may have elements of both.

The unitary state

The unitary state is built up around a single political centre and the territory of the state is under the control of this centre. Examples include Britain and France where political power is concentrated respectively in London (see Figure 5.5) and Paris. In chapter 1 the core–periphery model was used to explain the working of the world economy. The core–periphery model has also been used to explain the territorial division of political power and economic wealth at the level of the state (Rokkan and Urwin, 1982). The centre is the fulcrum of power in the state. It is the place where:

- powerful figures reside and meet
- it contains arenas for deliberations, negotiation and ritual ceremonies which affirm national identity
- it contains the symbols of power and national identity, e.g. 'national' buildings and monuments
- it contains the largest proportion of government information

The periphery, or to be more accurate, the peripheries, exercise limited power and look to the centre for key political decisions and resource allocation.

Both London and Paris are the largest cities in their respective countries, the most powerful economic and political leaders meet there and the two capitals contain the parliamentary assemblies and the bulk of the most powerful sections of the public sector. The two cities therefore contain the most potent symbols of national identity.

Figure 5.6 Federal system of the USA

The federal state

In the federal state there are a variety of power centres. This may be the result of the voluntary association of distinct territories. This *bottom-up system of federalism* can be found in the political history of the United States, Canada and Australia. In each of these countries individual states joined to form a federation of states making up one federal state (see Figure 5.6). It may also occur as a result of changes imposed from the centre or from outside. In this *top-down system* a large measure of control is often kept in specific parts of the territory.

We can identify strong and weak federations. A strong federal structure is one in which the constituent states are strong because they retain a large degree of power and the federal state is weak. In a weak federal system the constituent states have only a limited degree of power and the federal state is strong.

In both types there is a degree of tension

BOX L: TOO BIG? AND TOO SMALL?

Nationalism and the nation-state bind people together but also demarcate parts of the world from others and separate out people from one another. After three hundred years the nation-state may be coming to the end of its shelf-life, its rationale undermined, paradoxically, because it is both too small and too big: too small because sharp spatial demarcations are becoming irrelevant to the dynamics of a world economy of multinational corporations and global markets: too large because most governments and politicians are too distant from the everyday experience of ordinary people.

The nation-state is too small to deal with world pollution and too big to cope with neighbourhood waste disposal.

The nation-state has become a spatial anachronism, too small to have the necessary global consciousness and too large to be sensitive to the needs of localities.

There are moves away from nationalism. In one direction is the globalism which sees the need for world solutions to what are in reality world problems of war and peace, poverty and plenty, sustainability and ecological harmony. In the other direction is the localism which aims to solve local problems and avoid the easy option of only worrying about ecological issues in the abstract or only if they occur on the other side of the world.

The slogan *Think global, act local* is an attempt to join the two. In Europe we are beginning to see the geopolitical consequence of these trends. On the one hand there is the inexorable move towards European integration and the creation of a European super-state which will eventually override the sovereignty of existing nation-states. And on the other hand there is the disintegration of existing nation-states as smaller, more localized loyalties and allegiances begin to emerge. In Spain, for example, Andalusia, Castille and Catalonia have always existed in uneasy harmony. When the power shifts from Madrid to Brussels they no longer need to speak to one another. In Britain, the Scots have remained in the Union for the want of any other alternative. But when money is doled out from Brussels rather than London then there seems little point in staying with the English. Nation-states were always a compromise, a point of tension between competing interests. In the different world of the next century they may become an irrelevancy.

between the different centres of power. The constituent states are eager to retain their power and maintain a degree of independence. The federal state is ever eager to extend its power and influence. The federal state seeks to produce national policies while constituent states seek a measure of autonomy. The conflict is exacerbated if there are differences in political persuasion between the two levels of power. Take the case of the civil rights legislation in the USA in the 1960s (an example discussed in great detail in chapter 8). In many southern states racist groups held power while blacks were effectively marginalized and deprived the full exercise of their political rights. However, local action in association with federal legislation successfully challenged this entrenched power structure.

Conflict becomes even more apparent when language, religious or ethnic differences overlie the differences between state and state. In Canada, for example, the tension between the Quebec provincial government and the Federal Government is more pronounced because of the language and religious identity of Quebec compared to the rest of Canada.

In reality most states have a combination of unitary and federal elements. The United Kingdom, for example, is often presented as the textbook study of a unitary state. However, in Northern Ireland public policy is influenced by the Anglo-Irish agreement involving representatives of London and Dublin; Scotland and Wales have their own government departments and cabinet minister; the Welsh Office and Scottish Office play an important part in evolving and implementing public policy in the respective countries. If not quite a federal state then neither is it a single unitary system. Similarly, for all the talk of a federal arrangement in the USA, Washington acts as the single most important political centre and the federal state has enormous influence and impact on all parts of the country.

THE LOCAL STATE

So far we have used the term 'state' to refer to a single unit of analysis. In reality the state administers territory through a hierarchy of levels. There are various forms of regional government in the constitutional states of a federal structure (e.g. California and Texas in the USA). Below this is the local state. In this final section let us look at the local state in some detail. By way of examples we will examine the organization and functioning of local states in Britain and the USA.

The local state in Britain

The local state in Britain is a creature of central government; it was created by the central state and its powers are limited by central authority. The Local Government Acts of 1888 and 1894 laid down a two-tiered structure. In England the first tier consisted of sixty-two counties responsible for the broad range of local government services (roads, housing, education, police, etc.), eighty-seven county boroughs, which were basically large towns and cities, with the same responsibilities, and the twenty-eight metropolitan boroughs within the overall jurisdiction of the London County Council, responsible for such things as public health, libraries, housing and recreation. The second tier consisted of municipal boroughs (smaller towns), urban districts and rural districts, which all had minor responsibilities. This system lasted with minor changes and revisions for approximately eighty years until the 1972 reorganization which came into effect in 1974 (see Figure 5.7).

In England six metropolitan counties were made responsible for overall planning, transport, police and fire services. These counties were subdivided into districts, which were responsible for education, personal social services, housing, local planning and environmental health. The non-metropolitan areas, which consisted of thirty-nine counties, were made responsible for the same things as metropolitan counties, plus education and social services, and 296 districts were formed with the same powers as metropolitan districts minus control over social services and education. At the base level of the non-metropolitan areas in

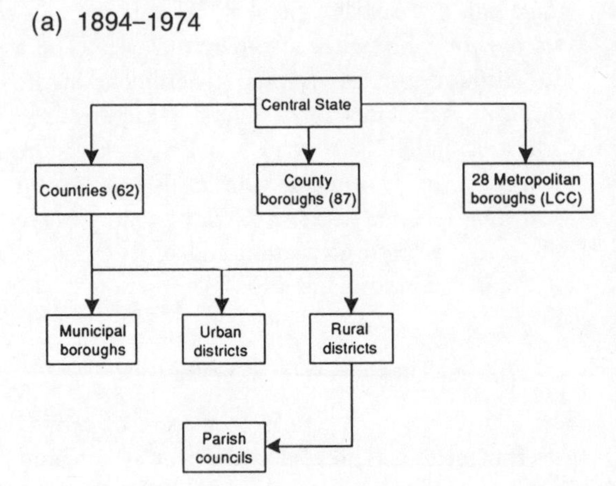

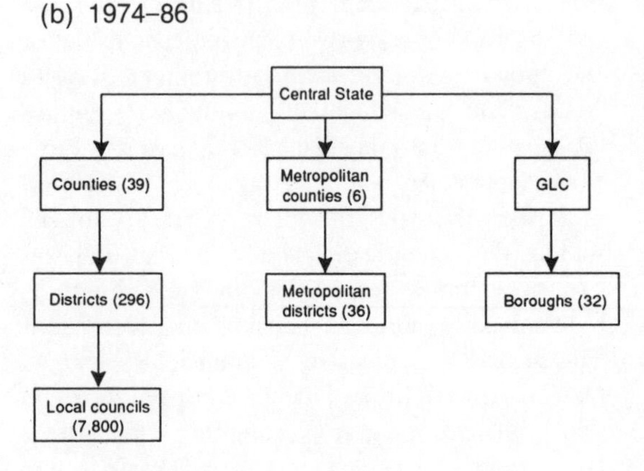

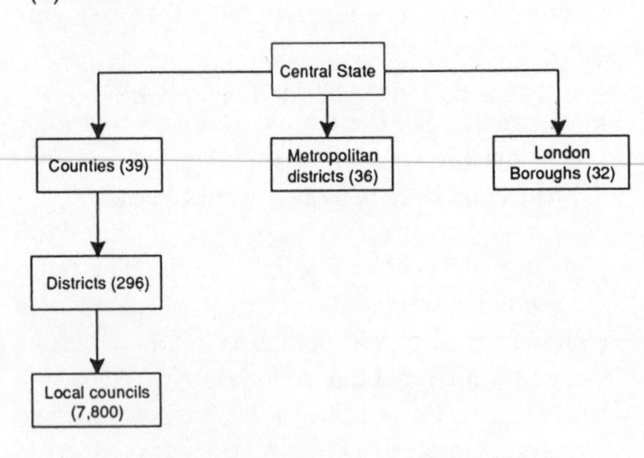

Figure 5.7 The local state in England

England are the 7800 local councils, responsible for the provision and maintenance of local amenities. The Greater London Council was made responsible for transport, overall planning and some housing, and the thirty-two London boroughs were in charge of housing, social services, leisure, public health and education (if outside the inner-city area).

The standard arguments put forward to explain reorganization centred around the concepts of size and efficiency. The existing authorities, so the standard and official argument goes, were too small, too inefficient and too fragmented for the expanded role of local government. Larger, fewer authorities, it was claimed, would cover the urban and suburban areas of city regions and would be more efficient in providing services. Some of the arguments were not borne out; before reorganization there were 141 planning authorities and after the much-vaunted reforms the number increased to 401. In reality, reorganization bore 'the imprint of political design' (Johnston, 1979, p. 150).

An alternative view of this local government reform has been advanced by Dearlove, who argues that such terms as 'efficiency' and 'rationalization' which are attached to the explanatory debates are meaningless; they are ahistorical, asocial categories which merely obfuscate the matter. He suggests that the real interests involved in reorganization need to be unpacked from the idealized terms of contemporary debates. He identifies two aspects:

1 The desire to increase efficiency, a central tenet of pro-reform arguments, stems from demands to reduce taxation in general and private-sector contributions in particular. The desire for efficiency, he argues, is really an expression of the demands to cut public expenditure in order to reduce company and individual taxes.
2 The enlargement of local authorities has been based on the attempt to make local government service more attractive to businessmen and executives, while the enlargement of city boundaries has been used to enfranchise the

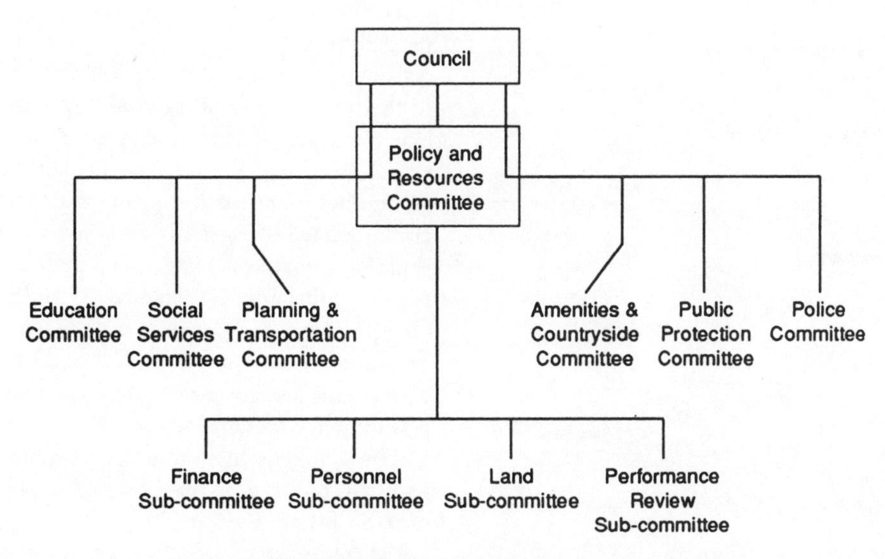

Figure 5.8 County committee structure

suburban middle classes so as to allow them to play a role in urban politics, a role that until recently was dominated by the power of the Labour Party. Dearlove summarizes reorganization thus: 'At best there is a romantic desire to capture some idealized form of conflict-free deferential democracy ... at worst an overt intent to rest local government power once again directly on top of economic power' (Dearlove, 1979, p. 105).

Local government was reorganized again in 1986 (see Figure 5.7c). This was part of an attack on local government by the Conservative government. The Conservatives came to power in 1979, committed to a reduction in welfare services and cutbacks in public expenditure. Prime targets were the big metropolitan counties and the Greater London Council. These were all Labour controlled in the early 1980s and constituted a point of resistance against the central power of a Conservative government. The role of place was also important. The headquarters of the GLC was right across the Thames from the Houses of Parliament and the GLC leader, Ken Livingston, was an impressive media performer. A Government White Paper published in 1983 entitled *Streamlining the Cities* argued that the abolition of the GLC and the metropolitan counties would reduce public expenditure and improve public services. The Government passed a Local Government rule in 1985 and the next year the GLC was abolished and the metropolitan counties were disbanded. It was a political act to achieve a political objective.

The local state in operation

The formal institution of authority in the local state is the council. It is elected on a broad franchise and everyone whose name appears on the electoral register is entitled to vote. Elections are held every four years on a ward basis. Each city is divided into wards whose populations range from 8000 in the smaller inner-city wards to over 30 000 in the more suburban areas. Full meetings of the council ratify major decisions but policy is formulated and policy implementation monitored at committee level; Figure 5.8 outlines a typical county committee structure. These committees rely on the local state bureaucrats for advice, information and guidance. The process of decision-making is represented in Figure 5.9. The chief officer's report on a particular matter goes to the various committees, whose recommendations are then used by the chief executive (the principal civil

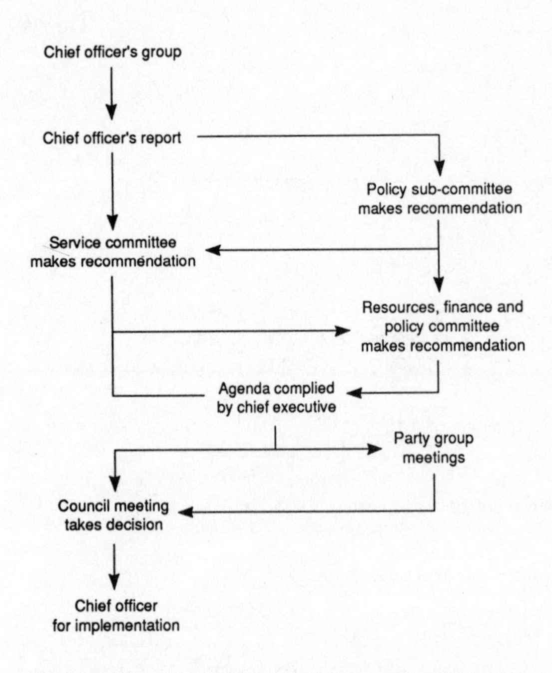

Figure 5.9 Formal decision-making in local government

servant) to compile an agenda. The points on the agenda are discussed and responses are formulated at the party group meetings held before committee meetings and full council meetings. The council rubber-stamps the decisions made by the majority party and these decisions are acted upon by the chief officers of the relevant departments.

The civil servants of the local state have significant influence. Their position as permanent experts, as initiators of local policy and as interpreters of central state directives gives them considerable power. They provide the information and the recommendations on which local politicians make decisions. The work of Saunders (1979) in Croydon and Newton (1977) in Birmingham points to the close relationship between chief officers and senior councillors, especially committee chairpersons. Newton found that the influence and experience of long-serving committee chairpersons served as a powerful counterweight to the professional expertise of the officials, and Saunders discovered that the relationship between chief officers and leading members of the majority political group was one of close allies. The political elite needed the advice and information from the officials, while the officials were obliged to take notice of the political and ideological dispositions of the elite members if their policy proposals were to stand any chance of being implemented. This second line of research suggests that rather than a crude councillors/officials dichotomy we need to think of a symbiotic relationship between the political elite of important councillors and committee chairpersons on the one hand and chief officers and chief executives on the other. Less powerful councillors and lower-level officials play a minor role in policy-making.

Local politics in Britain are dominated by the national political parties. National politics feed into local state policy-making. The party in power at the central state level attempts to implement its policies. Legislation passed by the central state often involves implementation by the local state. The policy guidelines from the central state are initially handled by the appropriate departments of the local authority, whose chief officer makes recommendations to the appropriate committee. The form of a local party's acceptance of policies emanating from the national party is mediated by the wishes of its constituents, especially those in marginal wards, and the politicking of various pressure groups. In general, the Labour Party is most receptive to the labour movement, while the Conservative Party is most receptive to pressure groups from the business community. On specific issues particular elements of these broad groups will be active. Building workers' unions and local builders, for example, will try to influence the level and form of local house-building plans. On other issues smaller, more peripheral organizations may play a role.

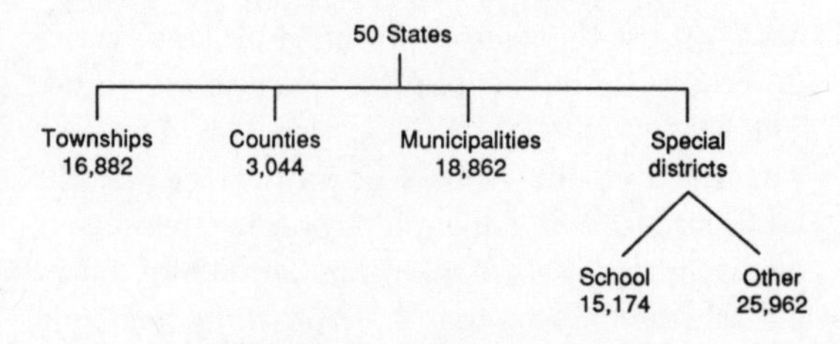

Figure 5.10 Local states in the USA

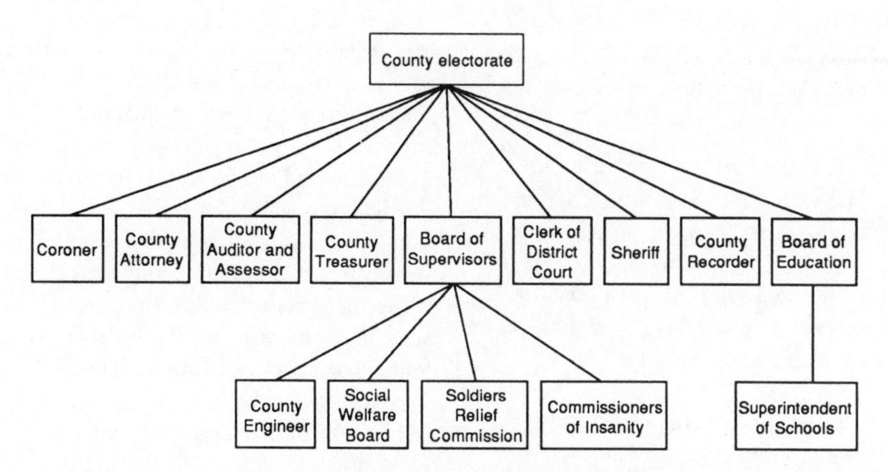

Figure 5.11 County government in Iowa

From the ingredients of national guidelines and directives, constituency demands and pressure-group activity, the local party in power will fashion its policies against the background of local political reality.

The local state in the USA

There are basically four types of local state in the USA: township, county, special district and municipality (see Figure 5.10).

Townships exist in only twenty-one states. In New England states they are the principal unit of local government. In the original townships decisions were made by a gathering of all adult males, who determined property rights and passed ordinances regarding public behaviour. In most places a more representative form of government has evolved; the electorate now votes for candidates for various public offices and for a three-person (or more) governing body.

Counties form the basic unit of local government below the state level and most of the USA is now covered by approximately 3000 counties. The functions performed by the county vary according to population density. Rural counties provide the basic services of maintaining law and order, providing road maintenance and supervising public health. Semi-rural counties fulfil these functions in addition to such services as library provision. Urban counties provide a wide range of services including street lighting, sewage disposal and garbage collection. Figure 5.11 outlines a typical situation, where the county electorate votes directly for various officers and the board of supervisors, who in turn direct the various boards and commissioners.

Special districts are the most numerous form of local authority in the USA. There are almost 16 000 school districts, and other types of special district total nearly 26 000. School districts have elected boards which have powers over the syllabus and teachers' salaries. Among the many types of *ad hoc* districts are fire service districts, park supervision districts, soil conservation districts, library districts, irrigation and drainage districts and even mosquito abatement districts.

Municipalities are the main form of government in urban areas. They are incorporated by the state in one of the following ways:

- under special charter a municipality is established by an act of the state legislature
- under a general law municipalities are created when a certain predetermined criterion is reached, e.g. if an area attains a certain population density or size
- under optional charter eligible residents may vote for a municipal government to be established

- under home rule measures eligible voters may submit proposals for municipal government to the state

The particular functions performed by municipalities are laid out in the respective charters. In general, municipalities provide police and fire protection, public works, libraries, parks, city planning measures and some low-cost housing.

To note the evolution of local government let us concentrate on the municipalities.

Municipal government

The form and functioning of municipal government in the USA have been shaped by the urban experience of the last 150 years and it is important to see the contemporary scene in the light of historical experience. In the pre-Civil War period cities were small and there was no large-scale established working class. Urban politics were dominated by an aristocratic oligarchy whose position was based on wealth and social standing. Conflict was muted. From the middle of the nineteenth century came the growth of a new merchant class in the expanding cities. The commercial elite displaced the old patrician class and the emphasis of urban politics under the new management was on civic promotion of economic growth through public investment and public works. In certain regions the merchant city gave way to the industrial city. Poor immigrants from Europe poured into cities where there was growing ethnic and class differentiation as the industrial revolution transformed the character of major US cities. With the demand for municipal services and the weight of the working-class vote came the boss system which was based on the control of votes. Through his organization the political boss controlled votes and votes gave power. The city boss used this power to maintain his position; he dispensed jobs, controlled elections to public office and held the city's purse-strings. From their position of strength the bosses channelled welfare services and jobs to certain groups and specific companies, who in turn voted and worked for the party machine. It was a mutually beneficial arrangement which enriched the boss and his supporters. It also fulfilled a number of other functions:

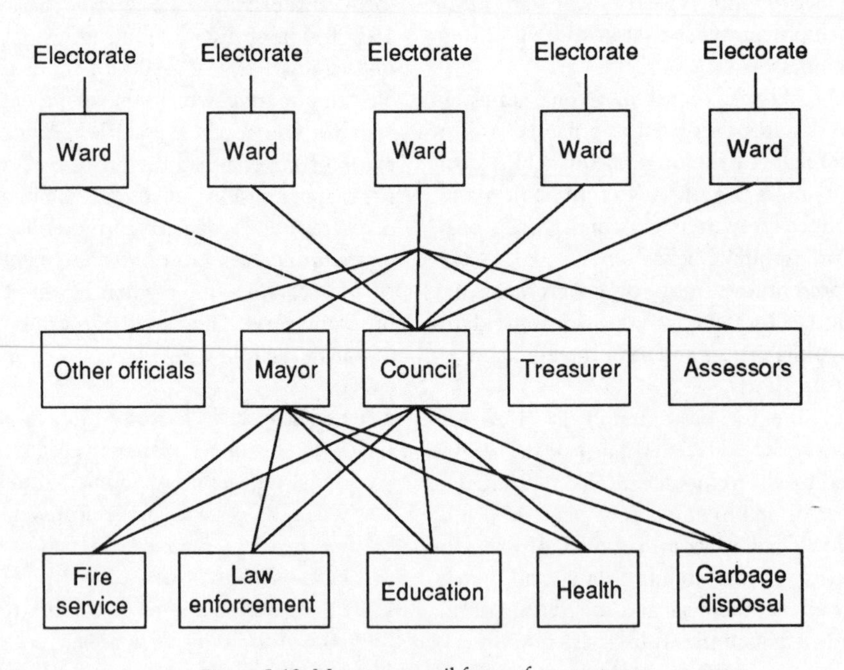

Figure 5.12 Mayor–council form of governance

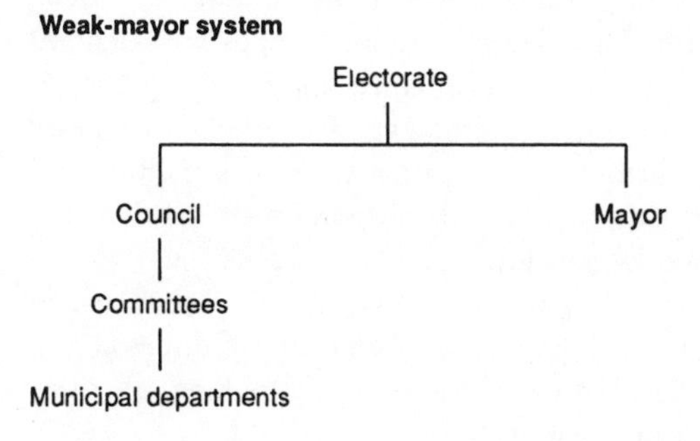

Figure 5.13 Alternative form of mayor–council relations

- It had an important socializing effect in so far as it provided immigrants (and the boss system was essentially the politics of the immigrant) with a political system which cushioned their arrival into a new environment.
- It was an inherently conservative system which legitimated the existing social and economic order because it did not mobilize the working class into a coherent group but exacerbated ethnic rivalries and promoted control and manipulation rather than democratic discussion. The boss system did not directly challenge the capitalist order.

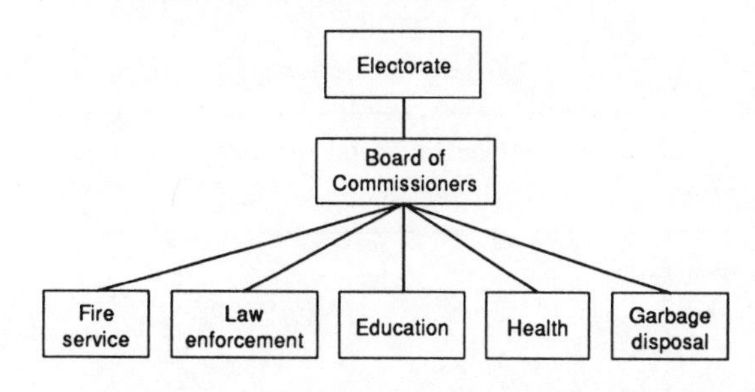

Figure 5.14 The commission form of governance

Conflict was reduced within the city by the ability of the upper- and middle-income groups to escape from the 'rabble-ruled' cities. The rich could move to the suburbs and then place barriers around their escape route. They escaped the influence of city politics by achieving political autonomy in suburban municipalities which conservative state legislatures safeguarded from central city annexation.

The boss system produced its critics. The reform movement started around 1900 and was led by business groups in each city. The problem, as they saw it, was that the boss system, while not directly challenging their economic interests, was not entirely satisfactory for business. It could become too expensive, it was unpredictable and in their nightmares they could see it as the 'mother and nurse of socialism'. The reform movement sought to keep business elites in political power by promoting non-partisan politics, the commission form of government and at-large elections. The various forms of contemporary city government reflect the differential diffusion and adoption of these reform measures.

There are basically three types of municipal government in the USA. The initial one, the one used by the bosses, was the *mayor–council* system (Figure 5.12). In this system the municipality is divided into wards represented by councillors who are elected for two to four years in office. The council is the main decision-making body and the mayor is the chief executive. Under the weak-mayor system (see Figure 5.13) the mayor is limited in the number of appointments he/she can make by the long-ballot system whereby city officials are directly elected, and by the direct supervision of municipal departments by council committees. Under the strong-mayor system the mayor exercises more power by his/her ability to appoint the chief administrators of the various departments, who are then responsible to the mayor rather than the council. Not too much should be made of this distinction in terms of differences in mayoral power. Chicago under Mayor Daly was theoretically a weak-mayor system: someone obviously forgot to tell Mayor Daly.

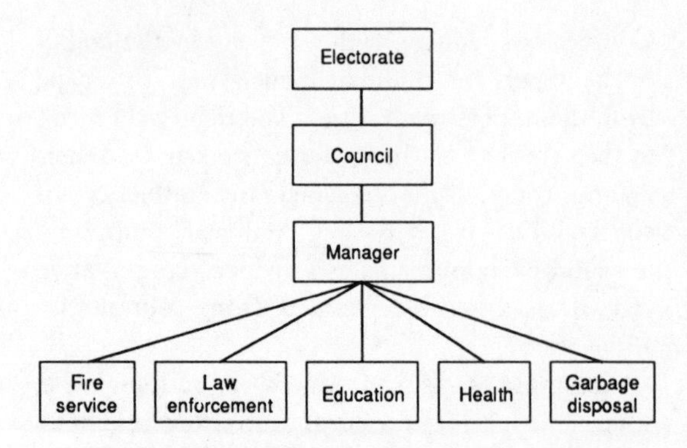

Figure 5.15 The council-manager form of governance

The majority of large urban municipalities operate the mayor–council system of government and over 80 per cent of all cities with a population greater than half a million operate in this way. It is the predominant system of municipal government in the large cities of the north-east and the mid-west of the country where the boss system was entrenched through the political power of the ethnic groups and the working class.

The *commission* form of government was first proposed by the reformers in 1900. Under this system the electorate votes through city-wide, non-partisan elections for a commission of three to seven members who direct municipal affairs for four years (Figure 5.14). The commission fixes tax rates, determines annual budgets and adopts general policies. This form of government has been especially important in the newer, smaller cities where the reformers had more power and there was no entrenched boss system. Only 10 per cent of municipalities operate this system.

The *council–manager* form of government was first proposed by the reformers in 1911. In this system the electorate votes, in non-partisan, city-wide elections, for a three- to nine-member council. The council sets taxes, determines budgets and hires a manager who is the chief executive directly supervising the municipal administration (Figure 5.15). The council–manager system became the preferred system of government for the reformers because it was not so time-consuming for the council as the commission form. The delegation of the day-to-day running of the administration to the manager made council office convenient for businessmen while the non-partisan ballot and city-wide elections diluted the power of ethnic minorities and low-income groups clustered in distinct residential areas. Under the council–manager system the discourse of political debate became dominated by the claims of efficiency and rationality rather than redistribution or social justice. This system is found in approximately 40 per cent of all municipalities and it is the predominant form of government in medium-sized cities with little or no experience of boss politics and where there is a high proportion of middle- and upper- income groups.

Having noted the example of local states in Britain and the USA, let us return to a discussion of the local state in general by looking at three important aspects:

- central–local relations
- influence of pressure groups
- areas of jurisdiction

Central–local relations

We can identify two extreme types. In the CENTRAL–local arrangement the central state retains most of the power and the local state has a subsidiary role. In the case of central–LOCAL the local state has most power. The former is more common than the latter. However, there is variation across this continuum over time and by different sets of public policy. The general trend has been for an increase in the power of the central state. As Table 5.1 shows with respect to the USA

Table 5.1 Government spending in the USA

	Percentage distribution of direct expenditure 1902	1980
Federal	34	51
State	8	20
Local	58	29
Total	100	100

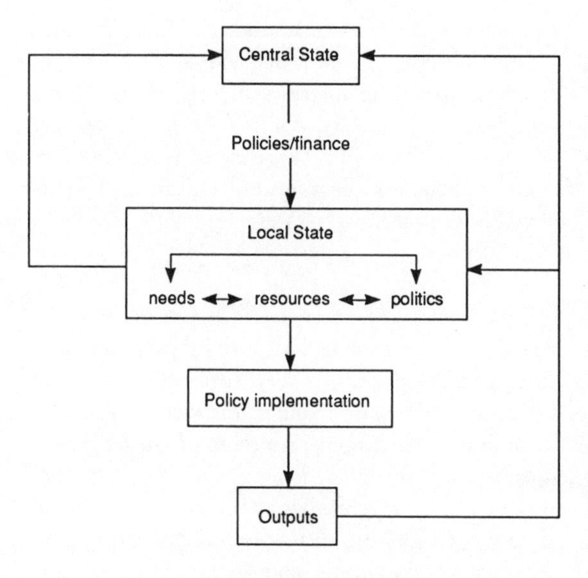

Figure 5.16 Central and local state

the power to spend money has moved towards the higher levels of government.

In the case of foreign policy and national defence most central states retain absolute power in deciding objectives. In the case of various welfare programmes, in contrast, such as housing, income support and education, the local state may play an important role. Ideally central governments want to keep control but pass responsibility onto the local state, which in turn wants both power and responsibility. If the central state keeps too much power it lacks the sensitivity of the local state to local conditions, but if it gives too much power to the local state this weakens its own position. The local state wants more independence from the central authorities but still wants central government to top up its local tax base. Too much independence and it is limited to the locally-generated tax base but too great a reliance on the central state reduces its power. The central–local state relationship is a tension between competing interests. In times of fiscal crisis the central state often attempts to pass the costs of the provision of public goods and services onto the local state level. Local states seek to resist this offloading.

The power of the local state varies in different countries. In Switzerland, for example, the local state has more power than occurs in either the USA or the UK. But even in these two countries the local state is more than just a neutral transmitter of central government policy.

Local states, even within individual countries, vary in how much they spend and how they spend it. The variation occurs because of differences in *demand*, *resources* and *dispositions* (see Figure 5.16).

Demand: There is a spatial variation in the demand (need) for the goods and services supplied by the local state, e.g. in poorer areas more people may require social welfare services while in richer areas there is a demand for more public open space. Local states vary in their type of constituencies.

Resources: Local states also vary in the resources at their disposal. Richer areas have a wider and deeper tax base than poor areas and can thus raise and hence spend more. Here we have an interesting paradox. The poorer areas may need public spending but have less ability to raise the money than the richer areas which can more easily raise the finance but have less need for it. This fiscal disparity problem may be overcome by revenue sharing schemes whereby central or federal governments top-up local resources. Where this does not occur there are real problems. Local states, like central governments, can experience a fiscal crisis as expenditure outruns revenues. The fiscal crisis of the local state is particularly acute in large cities with a high proportion of poor people and a declining tax base.

Disposition: Local states vary in policy formulation and implementation because of variations in political allegiance. In the UK, for example, the big urban authorities tend to be more radical than the rural counties. The result is that the urban authorities tend to spend more on public services than the rural areas. The administration of local states reflects the local political culture.

In this regard local states may also differ from the central authorities. Following on from the typology of functions and expenditure proposed by O'Connor (see Figure 4.5), Peter Saunders argues that there is a division of labour between the local and central state. The central state is concerned with ensuring capital accumulation and is most directly involved with social investment such as roads and various public works contracts. The local state is most directly concerned with the functions associated with legitimation and the social consumption and social expense expenditures (see Figure 4.3).

Saunders goes on to suggest that the difference between the central and local state in terms of function and type of expenditure is overlaid by the differences in decision-making. Social investment policies are made by the central state in association with corporate interests, with the emphasis on economic priorities and rational planning. The local state, in contrast, is relatively more open to democratic pressures. The argument is summarized thus:

> We must distinguish between social investment policies determined within the corporate sector at national and regional levels of government, and social consumption policies, determined through competitive political struggles often at local level. This means that the tension between economic and social priorities, between rational planning and democratic accountability and between centralized direction and local responsiveness, tend to underlie one another.
> (Saunders, 1980, p. 551)

In this light the local state is seen as the Achilles heel of the state apparatus. It is open to democratic pressure and it is concerned with the provision of services based on criteria of need rather than profit or ability to pay. The local state is a part of the state apparatus but it is a vulnerable part, suggests Saunders, a part which can be used to achieve real gains and defend real advances.

A distinction has to be drawn between the opportunities afforded by the local state in different types of society. In Western European countries to varying extents there is a socialist tradition often strongly represented in political parties operating at the local level. The local state in certain regions of the country can thus, holding everything else constant, act with some degree of freedom from the interests of capital. In North America, by contrast, there is a much weaker socialist tradition. The local state is much more open to business pressure and constrained in its social consumption expenses.

Influence of pressure groups

The decentralization of power to local areas makes the exercise of power very susceptible to local influences. We can make a distinction between local states which generate most of their revenue locally and those which receive a substantial proportion from central government. When a local state has to rely on local sources it is particularly sensitive to the needs and wishes of local tax-generating individuals and institutions. Local states are less sensitive when more funds come from central government. The difference can be seen in a rough comparison between local states in the UK, where traditionally over two-thirds of funds come from the central government, and the USA, where the majority of funds are raised locally. The differences are apparent in a comparison of the urban renewal schemes of the 1950s, 1960s and early 1970s which were undertaken by municipal authorities in both the UK and the USA. In both countries building contractors used their power to influence and encourage renewal. In the UK, however, new shopping centres and office buildings were constructed but so was public housing. The predominantly Labour-controlled councils sought to maintain their political support by meeting the housing needs of their constituents. In the USA, in contrast, there were markedly different redistributional consequences. The winners were big business and construction companies. The losers were low-income, predominantly black households who faced further restrictions on their already slight housing opportunities as cheap inner-city housing was demolished to make way for commercial developments and more expensive housing. The benefits of urban-renewal policies accrued to finance and construction capital and the costs were borne by low-income households in the inner city.

Because of the pressing need to generate income, the most influential pressure groups in US municipal politics are those representing business interests. They are not a homogeneous group and a distinction can be drawn between locally orientated smaller-scale businesses, downtown business interests, such as major banks and department stores which have major investment in the central areas of particular cities, and the large corporations operating through their branch plants. These business interests, acting sometimes together, sometimes alone, have been the single most important group influencing urban policies. The downtown interests have been especially successful in managing to get urban-renewal policies implemented. There has been a convergence of interests between business concerns and politicians. The politicians must meet, at least in part, the demands of the voters if they are to stay in power but they must also provide the best conditions for private investors and business if loans are to be raised and the tax base is to be maintained. Under present arrangements the politicians can only provide the municipal services which assure them of a power base by aligning their policies to the interests of capital.

Even in the UK, however, businesses are important actors in local politics. During economic downturns local states are competing for fewer tax- and employment-generating corporations. Local states are thus put in the position of vying with each other for a diminishing pool of business. The end result is that business interests can make high demands on local states, e.g. tax-free holiday periods, grants and loans. A spiral of demands can be established as local states compete to offer the best package to attract footloose industries and retain existing institutions within their jurisdiction.

There are pressure groups and pressure groups. The most successful are not pressure groups at all since their interests and outlook are shared by the political elite. Apart from those 'red islands' where left-wing Labour parties are in control, the local state in Britain, as in the USA, is suffused with the ideology of business interests. The interests of the 'general public', as perceived by most political elites, correspond with the interests of big business. The least successful pressure groups – the poor, the badly housed, the ill-educated and the unemployed – are also not real pressure groups, since their interests rarely appear on the political agendas. Between these two extremes there are a variety of groups whose success lies in their resources, their tactics, the stakes involved and the congruence of their interests with those of the political elite.

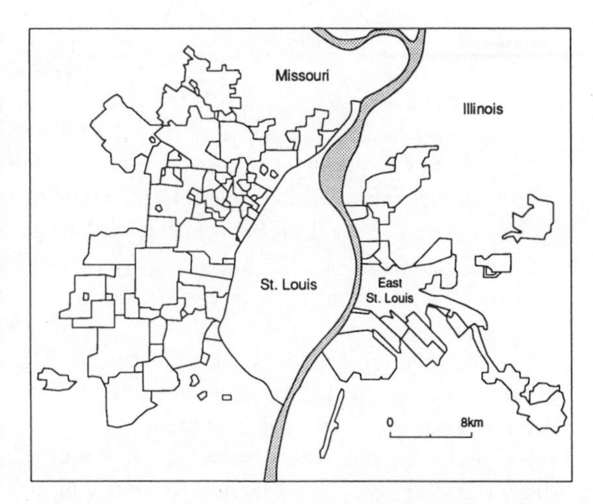

Figure 5.17 Jurisdictional fragmentation in the St Louis metropolitan area

Area of jurisdiction

Local states are responsible for particular pieces of territory. The boundaries of the local state are very porous compared to those of the state; economic activity and patterns of commuting cross and recross the boundaries of the local government more so than the state. This can pose a number of problems. Let us look at two in particular:

- the free rider
- metropolitan fragmentation

A *free rider* is someone who gets something for nothing. If two local states are close together, one provides free public open space facilities and the

BOX M: THE SECRET STATE

States have hidden agendas. Figure M.1 shows the command points and regional demarcation in the event of a 'national emergency'. Sub-regional HQs are to be given total power in the event of civil disturbance (within or without the context of nuclear war); their prime purpose is control. The people may be dead but the state will live on.

other does not, then residents of the other, lower-tax state can use the facility without paying. They get a free ride at the expense of someone else. Free rides occur because there may be an unequal distribution of costs and benefits. Benefits can be open to all within a large area but costs are borne by those within a particular local state. The solution lies in local state boundaries which match political and economic realities. There is resistance. Free riders want to maintain their free ride.

Let us examine the most exaggerated case of *metropolitan fragmentation*, the USA. Metropolitan areas in most of the USA comprise a central city hemmed in by suburban municipalities (see Figure 5.17). Prior to 1900 annexation of rural land by central cities was relatively easy and it was promoted by real-estate firms and construction companies. Since 1900 annexation has become more difficult as the rich have sought to defend their suburban peace and lower taxes. State legislatures have been used to stop central cities incorporating the suburban areas. Annexation still continues, especially in the expanding cities of the sunbelt, but there is marked conflict between the real-estate and construction firms eager for municipal governments to provide the essential services for their development projects and the residents keen to restrict urban encroachment which would affect their life-style and property values. In certain metropolitan regions the power lies with the developers, in others it rests with resident groups. The shifting balance of power is reflected in the pace and character of incorporation and annexation.

In the older cities political fragmentation has led to the *central city–suburban fiscal disparity problem*. This is quite a mouthful. What it means is this, the central cities have a relatively higher proportion of their population on welfare and greater demands placed on their public services, but only limited ability to finance these expenditures. In the USA the bulk of public service is provided by local taxes. The tax is based on the assessed value of all taxable property within the municipality. The problem for the central cities is a double bind: the dwindling tax base caused by the flight to the suburbs of higher-income households and industrial and commercial enterprises, and increasing expenditure.

The policy of declaring a fiscal crisis can be seen as an attempt to subdue popular demands and legitimate reductions in public services and municipal employment (Friedland *et al.*, 1977). In the suburbs, by contrast, there is a favourable fiscal position because of the relatively affluent population, the strong tax base and the few demands placed on municipal revenues.

The central city–suburban fiscal disparities are greatest in the older industrial regions of the north-east and mid-west, where there is a higher proportion of poorer households in cities, a weaker economic base and strong barriers to annexation and incorporation. The disparities are smallest in the expanding cities of the sunbelt, where the urban population is relatively richer, the industrial and commercial base is stronger and there are fewer barriers to annexation.

Metropolitan fragmentation overlies segregation of different income and ethnic groups. Poor blacks are concentrated in the central cities. This segregation is maintained because poorer inner-city residents are denied access to the suburbs by exclusive practices. Suburban municipalities zone

Figure M.1 Command points and regional governments in UK in the event of a 'national emergency'

minimum lot sizes or certain design standards for houses which effectively keep out lower-income house purchasers.

Metropolitan fragmentation increases social conflict. Within the central city the dwindling tax base and growing demands for public services raise conflict between the discontented taxpayers, who want their taxes reduced, the municipal workers (an increasingly important sector of total employment in many large cities) who want their incomes maintained and jobs secured, and the community groups, who want more and better services. The claims are incompatible and the unfolding tensions and conflicts between these groups and the city government raise the political temperature of the metropolitan core.

GUIDE TO FURTHER READING

On the nation-state see:

Anderson, B. (1983) *Imagined Communities: Reflections on the Origin and Spread of Nationalism*. Verso, London.

Chisholm, M. D. and Smith, D. M. (eds) (1990) *Shared Space, Divided Space. Essays on Conflict and Territorial Organization*. Unwin Hyman, London.

Gellner, E. (1983) *Nations and Nationalism*. Basil Blackwell, Oxford.

Hechter, M. (1975) *Internal Colonialism*. Routledge & Kegan Paul, London

Hobsbawm, E. J. (1990) *Nations and Nationalism since 1780: Programme, Myth and Reality*. Cambridge University Press, Cambridge.

Johnston, R. J., Knight, D. B. and Kofman, E. (eds) (1988) *Nationalism, self-determination and political geography*. Croom Helm, London.

Kedourie, E. (1985) *Nationalism*. Hutchinson, London.

Mellor, R. E. H. (1989) *Nations, State and Territory*. Routledge, London.

Smith, A. D. (1979) *Nationalism in the Twentieth Century*. Martin Robertson, Oxford.

Smith, A. D. (1988) *The Ethnic Origin of Nations*. Basil Blackwell, Oxford.

Tilly, C. (1975) *The Formation of National States in Western Europe*. Princeton University Press, Princeton.

Tivey, L. (1981) *The Nation-State*. Martin Robertson, Oxford.

On the spatial organization of the state consider:

Johnston, R. J. (1979) *Political, Electoral and Spatial Systems*. Oxford University Press, Oxford.

Johnston, R. J. (1982) *Geography and the State* Macmillan, London.

Paddison, R. (1983) *The Fragmented State: The Political Geography of Power*. Basil Blackwell, Oxford.

Rokkan, S. and Urwin, D. W. (eds) (1982) *Politics of Territorial Identity*. Sage, Beverly Hills and London.

On the local state see:

Boddy, M. and Fudge, C. (eds) (1984) *Local Socialism*. Macmillan, London.

Cockburn, C. (1977) *The Local State*. Pluto Press, London.

Cox, K. R. and Johnston, R. J. (eds) (1982) *Conflict, Politics and the Urban Scene*. Longman, London.

Duncan, S. S. and Goodwin, M. (1988) *The Local State and Uneven Development*. Polity Press, Cambridge.

Loughton, M. (1986) *Local Government in the Modern State*. Sweet & Maxwell, London.

Saunders, P. (1979) *Urban Politics*. Hutchinson, London.

Examples of the local state in operation include:

Bassett, K. A. (1990) 'Labour in the sunbelt: the politics of local economic development strategy in an 'M4 – corridor', town', *Political Geography Quarterly*, 9, 67–83.

Johnston, R. J. (1984) *Residential Segregation. The State and Constitutional Conflict in American Urban Areas*. Academic Press, London.

Newton, K. and Karran, T. (1985) *The Politics of Local Expenditure*. Macmillan, London.

Short, J. R., Fleming, S. and Witt, S. (1986) *House-building, Planning and Community Action: the Production and Negotiation of the Built Environment*. Routledge & Kegan Paul, London.

Relevant journals

Economic Geography
Environment and Planning (A, C and D)
International Journal of Urban and Regional Research
Local Government Studies
Policy and Politics
Political Geography Quarterly
Regional Studies

Other works cited in this chapter

Dearlove, J. (1979) *The Reorganization of British Local Government*. Cambridge University Press, Cambridge.

Friedland, R., Piven, F. F. and Alford, R. R. (1977) 'Political conflict, urban structure and the fiscal crisis', *International Journal of Urban and Regional Research*, 1, 447–71.

Gellner, E. (1964) *Thought and Change*. University of Chicago Press, Chicago.

Hechter, M. (1975) *Internal Colonisation*. Routledge & Kegan Paul, London.

Lichteim, G. (1974) *Imperialism*. Penguin, Harmondsworth.

Nairn, T. (1977) *The Break-up of Britain*. Nearleft Books, London.

Newton, K. (1977) *Second City Politics*. Clarendon Press, Oxford.

Pounds, N. J. G. (1972) *Political Geography* McGraw-Hill, New York.

Saunders, P. (1980) 'Local government and the state', *New Society*, 550–1.

6

THE STATE AS SPATIAL ENTITY

States are a spatial unit of the earth's surface. Their spatial quality is not a secondary feature, it is of major importance. A number of spatial relationships can be identified between:

- the people and the environment
- the state and the environment
- the people and the state

PEOPLE AND ENVIRONMENT

States occupy territory. In the process of nation-building this occupancy becomes the basis for a whole set of beliefs about the relationships between people and their environment. I will use the term *national environmental ideology* to refer to this set of beliefs. Let us examine each of these terms:

National in the sense that a whole set of myths and beliefs are established in the process of nation-building. For example, the creation of the USA involved subduing 'wilderness' and the notion of extending the frontier is an important element of US national identity.

Ideology in the sense of a partial set of beliefs which highlight the experience of some groups and ignore or marginalize the experience of others. For example, up until very recently the history of the American west ignored the role of blacks, women and the plight of the 'Indians'. It was a history of those moving west, not of those facing east.

The *environmental* element can be broken into the elements of particular places and general spaces. Every country has a set of symbolic places which condense popular feelings of nationhood. There are the historic sites, e.g. Gettysburg (USA), which are important in the historical evolution of the country. There are also the special sites such as the White House (USA), Parliament Building (Australia), or the Houses of Parliament (UK) which function as symbols of statehood. Particular places in the space of the state are given importance and priority as recorders, containers or reflections of national identity.

There are also more general attitudes to the space of the country. Three types of general space can be identified:

- wilderness
- countryside
- city

Wilderness: Up until the twentieth century the dominant view in most states was that the 'defeat' of the 'wilderness' was a sign of progress to be encouraged and celebrated. Governments in the New World actually sought to 'open up' the 'wilderness' by defeating the indigenous population, cultivating the land and building towns and cities. In nineteenth-century USA, for example, the drive westward was seen as a symbol of strength and unity, the extension of the frontier further westward was a sign of national progress.

In the late nineteenth and, particularly, in the twentieth century an alternative view has emerged. This view sees the wilderness as the authentic landscape whose destruction is a source of regret and unease. The extension of the frontier and the defeat of the wilderness is a sign of regress not progress.

These differing views can catch the public imagination and, through the power of public opinion, influence government action. In the richer countries a balance has to be struck between utilizing the economic resource of the territory and preserving remnants of wilderness. The production and conservation of particular pieces of territory becomes an element of government policy, sometimes in competition with the full development of the economic potential of the resource base of the country. In the poorer countries of the world the balance of political power is often weighed more heavily in favour of development rather than conservation. However, world opinion can affect national policies. The debate about preserving the tropical rain forest is a case in point. Foreign opinion can act as a counter to the demands of the big corporations which seek to exploit wilderness areas.

Countryside: In English the term country has a double meaning: it is used to refer to rural areas; it is also used as another word for the nation. This double usage reflects the huge importance that rural images play in the creation and recreation of national identity. Take the case of England: one of its most important icons is the image of a green and pleasant land, a land of small villages and green fields. In England, as in many other countries of the world, the rural landscape provides a rich source of images of national identity.

The cultural importance of the rural is also reflected in political debates. In many rich countries there is a high level of subsidy to farmers, often in contrast to low levels given to industry. This subsidization reflects the political power of the agricultural lobby, but it also represents the feeling that farming and farmers have a special place. There is also the argument of agricultural fundamentalism which sees farming as the bedrock of all economic activity, a doctrine first formulated in the eighteenth century but trotted out ever since whenever more than two farmers meet. Crises in farming are seen as events of major importance, affecting the very heartland of a nation.

There are also debates about the look of the countryside. In the 1970s and 1980s a number of writers in Britain complained that the growing mechanization of agriculture and the element of subsidies was leading to a destruction of the traditional countryside (e.g. Shoard, 1987). The argument obtained popular approval because there was a very strong feeling that the countryside was a national asset. Farming land was private property but also part of the national heritage. These debates reflected the importance of environmental imagery in political debates.

City: There are two contrasting beliefs about the city. On the one hand, the city is contrasted unfavourably with the countryside as a place of vice and debauchery, almost a foreign land where community ties are eroded. This is the city as Babylon. On the other hand there is the view of the city as a place of freedom, a site for innovation, a centre for radicalism, compared to the stuffy conservatism of the countryside. This is the city as Jerusalem.

These contrasting images influence public policies. The city as Babylon is used to legitimate policies of public spending and political representation which favour rural areas at the expense of cities. Big city building programmes, as in the British new town scheme, or the grand urban renewal schemes of President Mitterand in Paris are justified with reference to the notion of the city as Jerusalem, a place for building a new and better society.

There is a complex relationship between national environmental ideologies and government policies. The ideologies are used to justify policies while specific policies may maintain the basis of the ideology. Consider farm support schemes. These are used to subsidize the farming sector. They are legitimized with reference to the ideology of agricultural fundamentalism and the cultural primacy of the rural. The policy of income support in turn helps to maintain the importance of the rural sector.

National identity is bound up with particular pieces of territory. National environmental ideologies are the territorial belief system which give coherence and meaning to a nation-state's view of itself. These ideologies are not static, they are points of debate and argument. For example, one part of the women's movement in the USA was to reclaim their historic role in the transformation of the wilderness. A number of feminist histories such as Kolodny (1975) pointed to the importance of women in the frontier. This historical re-examination was part of the struggle for contemporary freedom. National environmental ideologies are sites of struggle.

THE STATE AND THE ENVIRONMENT

States have responsibility for particular pieces of territory. The nature of this environmental legacy can influence the power of the state. Let us consider:

- resource endowment
- size and location

Resource endowment

Some states are luckier than others. Their territory contains resources – e.g. oil reserves, mineral deposits, rich soil – which can be turned into commodities and hence provide a strong fiscal base for the state. A continuum can be drawn from resource-rich countries at one end to resource-poor countries at the other.

Resource endowments are not static. They vary according to a whole set of variables. Let us look at just three:

- levels of technology
- market conditions
- consumer preferences

Levels of technology: Uranium has always been located in the Northern Territory of Australia. For at least 50 million years it was simply something in the ground, in places Aborigines avoided because they knew it made people ill. Since the 1950s, however, because of the development of the nuclear industry, it has become a valuable resource, highly-prized in world markets. The 'move' from simply a piece of ground to valuable resource is because of technological changes and the perception that bauxite is a resource.

Market conditions: Commodities can be made more valuable through control of the market. In the 1960s oil was a commodity and countries which had such deposits were fortunate. Fortunate but not wealthy. In the Third World, exploitation of the oil fields was in the hands of the multinational companies. When states began to take a more direct role, oil became more valuable and, after 1973 and the OPEC decision to raise the price, it became even more valuable. Oil producing countries had increased the value of their resource.

Consumer preferences: In the last century and for almost two-thirds of this century, many of the coastal areas of the Mediterranean were thinly populated. With the rise of disposable income, more holidays and consumer preferences which valued beach holidays, the Mediterranean coast saw a huge building boom. For countries such as Spain foreign tourism became a major money-earner. Beaches which for years had seen little more than a few fishing boats now became packed with northern Europeans seeking the sun. The coastal zone and the hot sunshine had become valuable resources.

The world trading system is one of power and negotiation in which states seek to maintain the advantage of their resource endowment. The power of a state is partly dependent on its position in the resource endowment cycle. In the eighteenth and nineteenth centuries, for example, Britain's power was based partly on its resource endowment of coal and iron-ore, the raw materials of an industrial economy and military power. We can think of cycles of resource endowment influencing state power and national standing. We should be wary of a deterministic argument which reads off a country's wealth from its resource endowment. Switzerland and Japan, to take the most obvious examples, have few mineral resources yet manage to be two of the richest countries in the world. The ultimate resource of any country is the population of that country, its vitality, inventiveness and imagination.

Size and location

The size of a country is important. Size can be measured in a number of ways:

- physical size
- population size
- economic size

Physical size: The larger a state the more chance of having a range of commodities. States such as the USA which straddle many degrees of latitude and longitude have the advantage of a variety of environmental conditions which can be utilized at different times in the resource cycle, e.g. the coal and iron-ore of the industrial north-east in the nineteenth and early twentieth centuries, the sunshine and climate of the sunbelt in the post-industrial phase. Large countries have a number of possible advantages from range of climates to variety of resources. There are also disadvantages. Big states may have difficulty in maintaining internal cohesion. Separatist movements flourish when there is distance between the centre and the periphery – physical distance as well as socio-economic distance. The two distances are often interrelated. Big states may therefore need some kind of federal solution but this may make difficult the even spread of national policies and objectives. Like all physical attributes, size can create problems as well as opportunities.

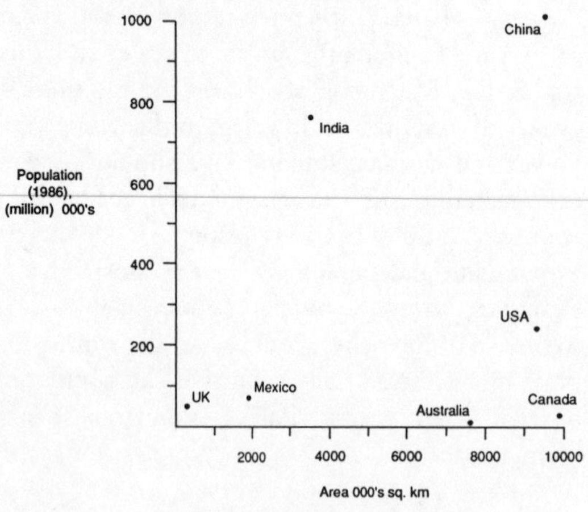

Figure 6.1 Physical and population size of selected countries

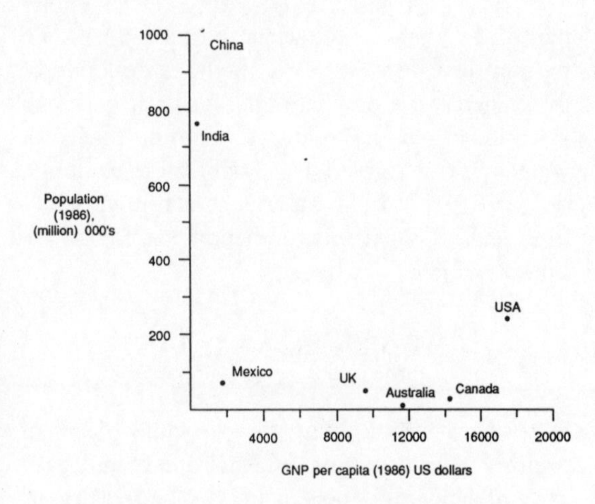

Figure 6.2 Population and economic size

Population size: The size of a country can alsc be measured with respect to population. Canada has a large surface area but a relatively small population. India, in contrast, has a smaller territory but a very much larger population (see Figure 6.1). Population size, like physical size, can be a source of strife as well as opportunity. Too many people, with respect to the ability of the economy to sustain them, can lead to such severe problems as endemic mass unemployment, with the associated tendency toward political instability. Too few people, and the pool of creative talent as well as the tax base is considerably reduced. Is there an optimum population? No simple answer can be given because it depends on the resources of the country and the wealth of the economy.

Economic size: Of major significance is the economic size of a country. Consider Brunei, a small country with a small population. However, its rich resource endowment of oil makes it one of the wealthiest single states in the world. Size of fiscal resources is an important indication in the ranking of states. Economic size can also be measured with respect to the market power of a nation. In this regard the United States is the most important nation in the world because its large, relatively affluent population constitutes a major economic force, the single biggest market and a pool of purchasing power second to none. China

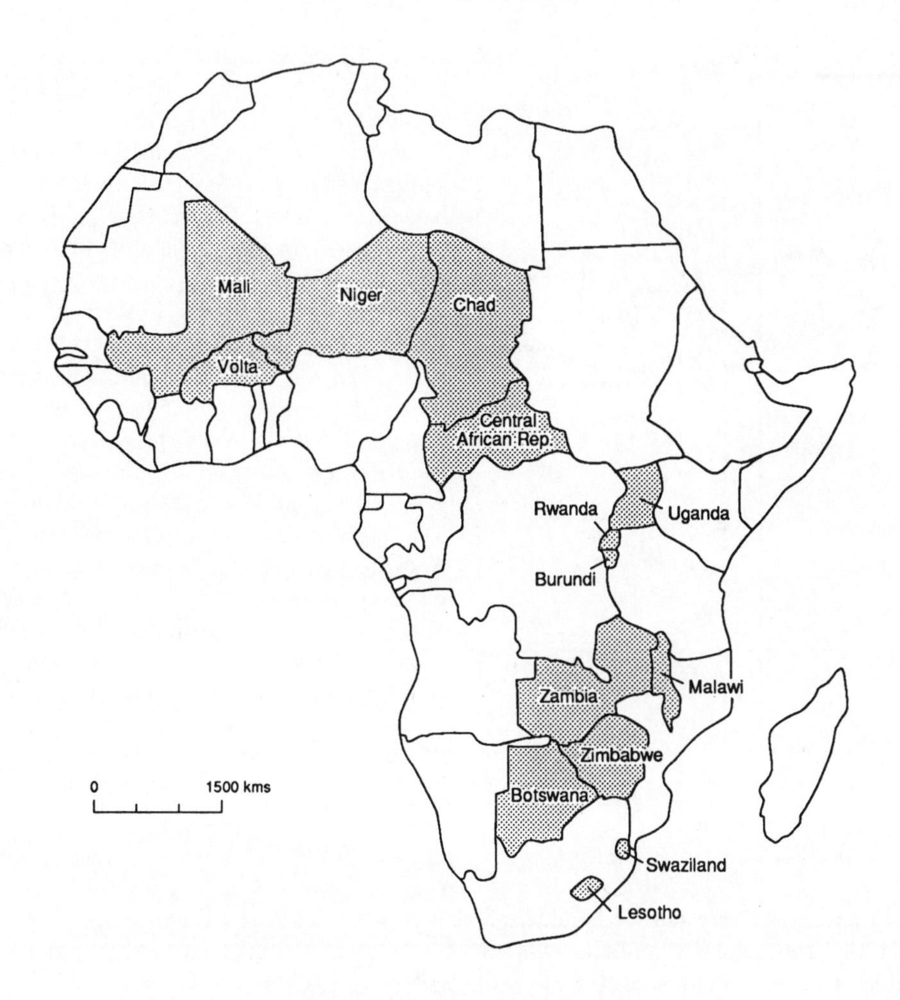

Figure 6.3 Landlocked states of Africa

has more people and a larger territory but has only a twentieth of the average US income (see Figure 6.2).

The *absolute location* of a state is important. We can, for example, distinguish between landlocked states and those states which have access to the sea. Landlocked states include: Bolivia, Paraguay, Central African Republic, Nepal, Czechoslovakia, Switzerland. These states, as the name implies, are surrounded on all sides by land (see Figure 6.3). They do not have independent access to relatively cheaper sea transportation and they are denied access to the rich resources of the ocean. But being landlocked does not necessarily mean being poorer. Switzerland is landlocked but is still one of the richest countries in the world.

The *relative location* of a state is another important dimension, i.e. relative in relation to other states. The point of contact is international boundaries. We can distinguish between so-called natural boundaries, such as mountain ranges, rivers and seas, and the more artificial boundaries of straight lines drawn on a map. I use the term 'so-called' natural because there is nothing natural about international boundaries, they are all human creations, although some are more obvious than others.

A further distinction can be made by the

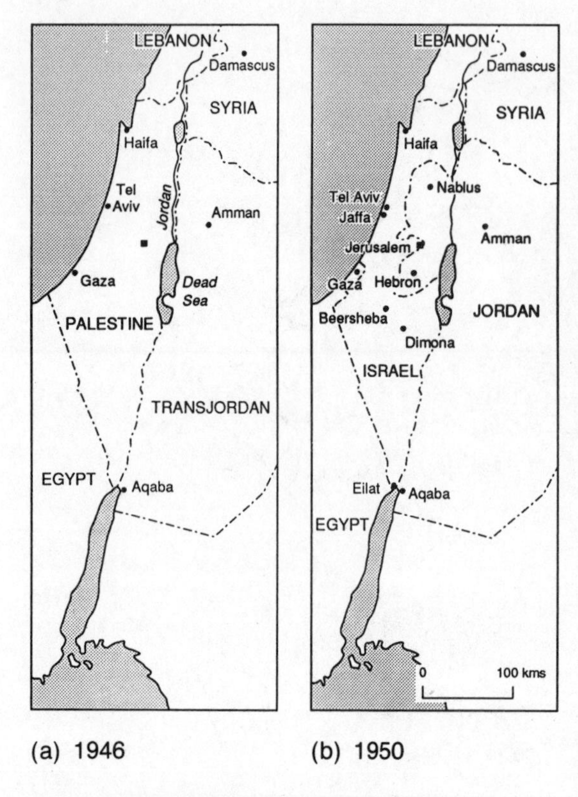

Figure 6.4 The creation of new states

porosity of frontiers. In very porous boundaries, e.g. Canada–USA, there are few limitations placed on the traffic of people, goods and ideas. Impervious boundaries, in contrast, pose a barrier between states, they reduce human movement and commercial trade. Boundaries are the point of contact between states. The nature of the relationships between states is often expressed through their frontier contact. When things are going well frontiers become more porous. When things are going badly, barriers go up and interaction is lessened. In such cases frontier landscapes may emerge on either side of the boundary, even landscapes of military presence.

Boundaries themselves can become a source of tension between states. Grundy-Warr and Schofield (1990) suggest that of the 300 land boundaries of the world at least 10 per cent are either undelimited or actively disputed. Iran–Iraq, Russia–China and India–Pakistan are just some of the pairs of countries whose boundaries are disputed.

We can also distinguish between *stable* and *unstable* boundaries. Stable boundaries, such as between Spain and France, have a long history of permanence. Unstable boundaries exist in places of rapid geopolitical change. We can exemplify this statement with references to Israel. After the First World War the Middle East was divided up by the imperial powers of France and Britain. Figure 6.4(a) shows the position in 1946. At this time the League of Nations Mandate (see chapter 2) gave control of Palestine and Transjordan to Britain. In 1947 Britain announced it would withdraw. There were then about 600 000 Jews and over a million Arabs in Palestine. The United Nations approved the creation of separate Jewish and Arab states. There was world support for the creation of a Jewish homeland both from Jews throughout the world and non-Jews eager to make amends for the

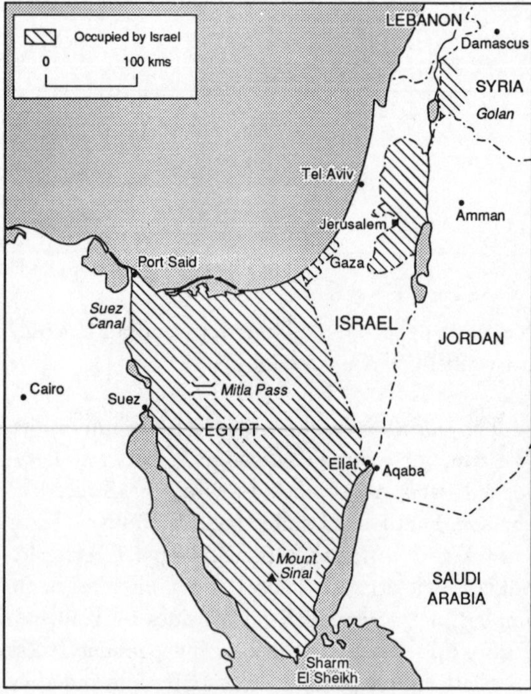

Figure 6.5 Changes after the 1967 War

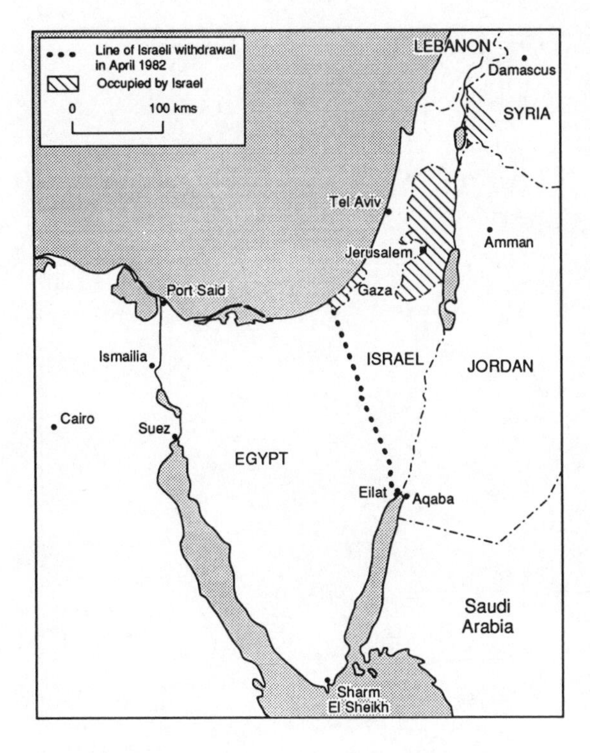

Figure 6.6 The position in 1990

dreadful wrong of the Holocaust. It prompted armed struggle between the two main groups in Palestine, the Arabs feeling they were losing land, the Jews eager to obtain control of the Promised Land. In 1948 the State of Israel was declared. It was immediately attacked by Arab armies. A truce was secured and the boundaries were stabilized (Figure 6.4 (b)). In 1967, the Six Day War broke out. Israel captured the Gaza Strip and the Sinai (from Egypt) the West Bank (from Jordon) and the Golan Heights (from Syria) – see Figure 6.5. Relations between Israel and Egypt subsequently improved and a settlement – the Camp David Treaty of 1979 – meant an Israeli withdrawal from the Sinai (see Figure 6.6). However, Israel still kept control of Gaza and the Golan Heights while continued control of East Jerusalem and the West Bank, with consequent Jewish settlements, caused deep resentment amongst the Arab population of these areas.

The constant boundary changes between Israel and surrounding areas is unusually unstable. However, they highlight, in a very dramatic way, how boundary changes reflect and embody the changing relations between states.

PEOPLE AND THE STATE

We can consider two aspects of the spatial relationship between people and the state:

- geography of elections
- geography of spending

The geography of elections

In democratic systems governments are voted in by the people. Elections are about turning votes into representation. The geography of elections is an important variable in determining political outcomes. We can identify three important elements:

1 the geography of voting
2 the geography of representation
3 the geography of electoral systems

The geography of voting

Not everyone votes the same way. There is a spatial variation to voting patterns. A hierarchy of differences can be identified.

National cleavages lead to major differences in political expression and hence in voting. Examples of such cleavages include urban–rural, dominant–dominated cultures and core–periphery regional economies. Throughout the twentieth century in the core countries of the world the major process has been the incorporation of peripheral representation into national representation. In the USA, for example, political expression at the federal level is essentially a two-party race between Democrats and Republicans. The process has not been so complete everywhere else. In Britain, for example the 1970s saw the rise of such peripheral groups as Plaid Cymru in Wales and the Scottish Nationalists. National cleavages affect the regional distribution of political representation and hence the regional geography of voting.

Neighbourhood effects: There are also more localized spatial effects. In the case of the neighbourhood effect voters are influenced by the local majority party. Let us illustrate with an example. Table 6.1 shows data for voting patterns disaggregated by type of constituency and perceived class of respondent. In the mining constituency, which was a Labour stronghold, 91 per cent of those seeing themselves as working class voted Labour and only 9 per cent voted Conservative. However, even 36 per cent of those classifying themselves as middle class voted Labour. Compare these figures with the voting pattern in the more Conservative-orientated resort constituency. Here the proportions voting Conservative, even amongst working-class people, increased. In other words, in Labour strongholds more people vote Labour irrespective of class, and vice versa for Conservative constituencies. The neighbourhood effect modifies the class variable.

Friends and neighbours: A related effect is the power of local candidates to affect voters in their home area. If we examine Figure 6.7 we can see that rival candidates for Republican nomination for the governorship of Vermont polled most votes in areas closest to their homes. In their immediate home environment candidates may be better known to local voters. Voters may also believe that the candidate understands the problems of the local area and, if electorally successful, will have an added incentive for doing something about solving them.

In reality national cleavages, neighbourhood effects and the friends and neighbours effect all come into play with local issues. For example, Figure 6.8 shows the electoral support for Governor George Wallace; notice how it is concentrated in the southern states. This is partly a function of his support base: he was Governor of Alabama and he was a distinctly southern candidate, electioneering on resistance to Civil Rights legislation. The concentration of his support reflected national cleavages, regional issues and the local concerns of white voters.

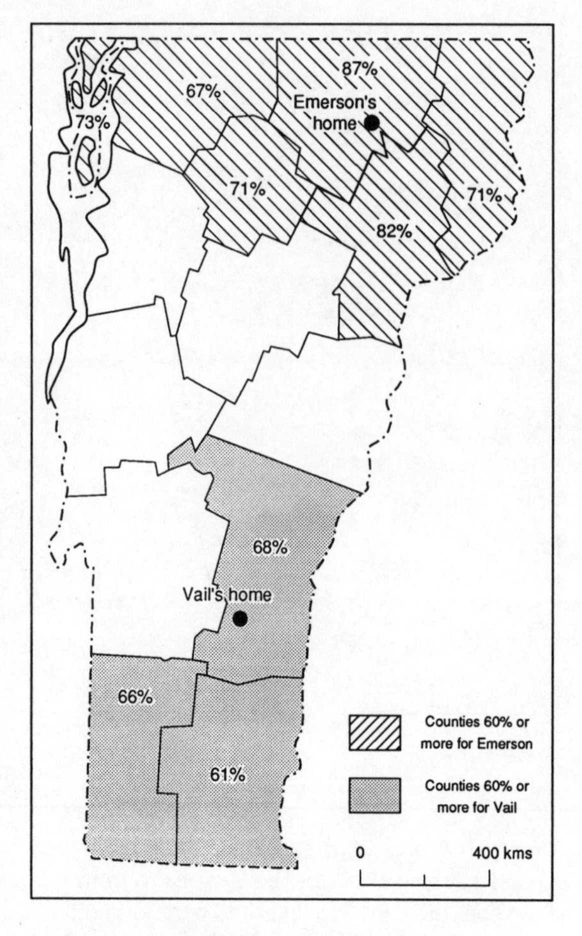

Figure 6.7 Voting for Republican nomination for governor, Vermont, 1952

The geography of representation

National territories are partitioned up into political constituencies. This is the geography of representation. Different partitions may produce different results even when the votes remain the same. Let us illustrate this remark: Figure 6.9 shows a simple model of two competing political parties, the Reds (R) and the Blues (B). The numbers refer to the votes cast for each party. In the case of Fig 6.9(a) there are two constituencies and the result, under a first-past-the-post system, is that each party wins one constituency. In the case of Figure 6.9(b), in contrast, even though each party received the same

Table 6.1 The neighbourhood effect

Type of constituency	*Mining town*		*Resort town*	
Perceived class of respondent	*Middle*	*Working*	*Middle*	*Working*
Per cent Voting:				
Conservative	64	9	93	52
Labour	36	91	7	48
Total	100	100	100	100

Source: Butler and Stokes, 1969, p. 183

number of total votes the nature of electoral constituencies leads to the Blues having a 3 to 1 majority. This phenomenon is known as *malapportionment.* It refers to the imbalance between the number of voters in each constituency. Under the four constituency system of Figure 6.9(b), the western constituency had 100 votes while the three eastern areas had no more than 34 votes. As this example shows, malapportionment can ensure electoral victory for one political party even though it may not have the majority of total votes.

There are many examples of malapportionment. Taylor and Johnston (1979) discuss, in some detail, the blatant malapportionment in nineteenth-century Britain and the USA before the

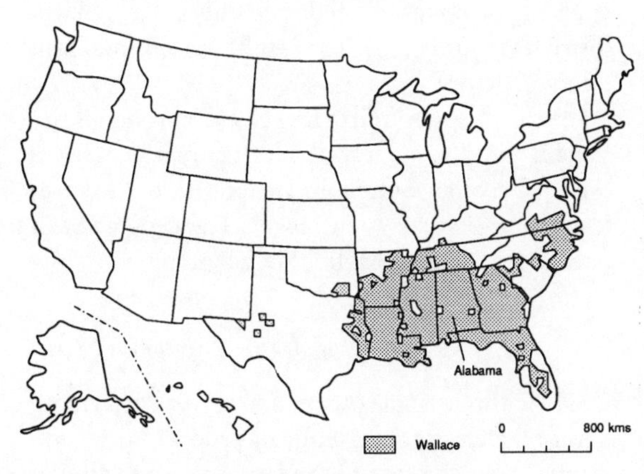

Figure 6.8 Majority electoral support for George Wallace in 1968 Presidential election

(a) Two constituency system

R 60 *B 40*	R 40 *B 60*

(b) Four constituency system

R 60 *B 40*	R 13 *B 20*
	R 13 *B 20*
	R 14 *B 20*

Figure 6.9 A simple model of voting

reapportionment revolution of the 1960s. In 1962 the Supreme Court of the USA ruled that malapportionment issues were a concern of the courts. Subsequent verdicts stated that population equality was to be the primary criterion for electoral boundary making in state legislative bodies and for the House of Representatives. The result was a reapportionment revolution. Some of the effects are shown in Table 6.2.

Malapportionment is maintained by political power. In most cases rural areas tend to be over-represented compared to urban areas. Since the rural areas have the majority they are unlikely to radically alter the system. This is especially true when urban and rural areas have different political persuasions. A good example is Queensland in Australia where rural over-representation was built into the electoral system.

This pattern become entrenched in Queensland because the ruling County Party, later the National

Table 6.2 Judicially inspired Congressional reapportionment

State	*Population ratio of largest to smallest constituency*	
	1962	*1965*
Texas	4.4	1.2
Michigan	4.5	1.0
Colorado	3.3	1.2
Maryland	3.2	1.3
Ohio	3.1	1.3
Georgia	3.0	1.4

Source: Baker, 1966

Party, gained from the existing malapportionment and saw no reason to change the system. It only needed 36 per cent of the votes to stay in power. Even though the metropolitan areas increased in population their electoral representation did not increase. The National Party, serving rural interests, held onto power in a state that was overwhelmingly urban. It only lost power in December 1989 when a bribery and corruption scandal meant that even the unfair voting system could not save it.

In some instances malapportionment can benefit urban areas. In post-war UK, for example, electoral reform was institutionalized in the 1949 Redistribution of Seats Act, which established the Boundary Commissioners in order to review constituencies. They met infrequently, with the result that the 1970 elections was fought within boundaries established in 1954. During that time population redistribution was one of suburbanization. Inner city constituencies lost population while suburban areas gained. As Table 6.3 shows, the result was over-representation of such inner-city areas as Birmingham Ladywood and Glasgow Kelvingrove, while suburban areas like Cheadle and Billericay were under-represented.

Table 6.3 Disparities in constituency population size, 1955–70

	1955	*1970*
Birmingham Ladywood	46904	18729
Glasgow Kelvingrove	39672	18907
Cheadle	61626	107225
Billericay	58872	123121

Even when constituencies are the same size, electoral boundaries can still influence the outcome of elections. The manipulation of boundaries to achieve a particular result is known as *gerrymandering*. The name owes its origins to Elbridge Gerry who was elected Governor of Massachusetts in 1810. In order to keep his party (the Republican-Democrats) in power he signed a bill drawing boundaries for political districts which favoured his party. It was a success. In the next election the Federalists won more votes, 51766 to 50164 but the Republican-Democrats won 29 Senate seats to their opponents' 11. The outline of the boundaries was described by a graphic artist as similar to the skeletal outline of a salamander. Adding the Governor's name gives us gerrymander.

Gerrymandering is the manipulation of political boundaries to achieve particular electoral outcomes. It happens all the time. Between 1958 and 1962 the Iowan Congressional representation was reduced by one. The Republicans were in charge of the legislature and redrew the boundaries so that the big-town Democrat support of Fort Dodge and Des Moines was packed into one new district and the remaining Democrat support was cracked amongst predominantly Republican districts (Figure 6.10). The result was a Republican election victory.

If the responsibility for electoral boundaries remains in the hands of elected representatives then it becomes one more weapon in the fight to keep control and stay in power. The geography of politics becomes the politics of geography.

The geography of electoral systems

Even without malapportionment or gerrymandering the electoral system can produce some odd results. Take the case of the UK general election of 1987. This was based, as in all previous elections of the twentieth century, on a first-past-the-post

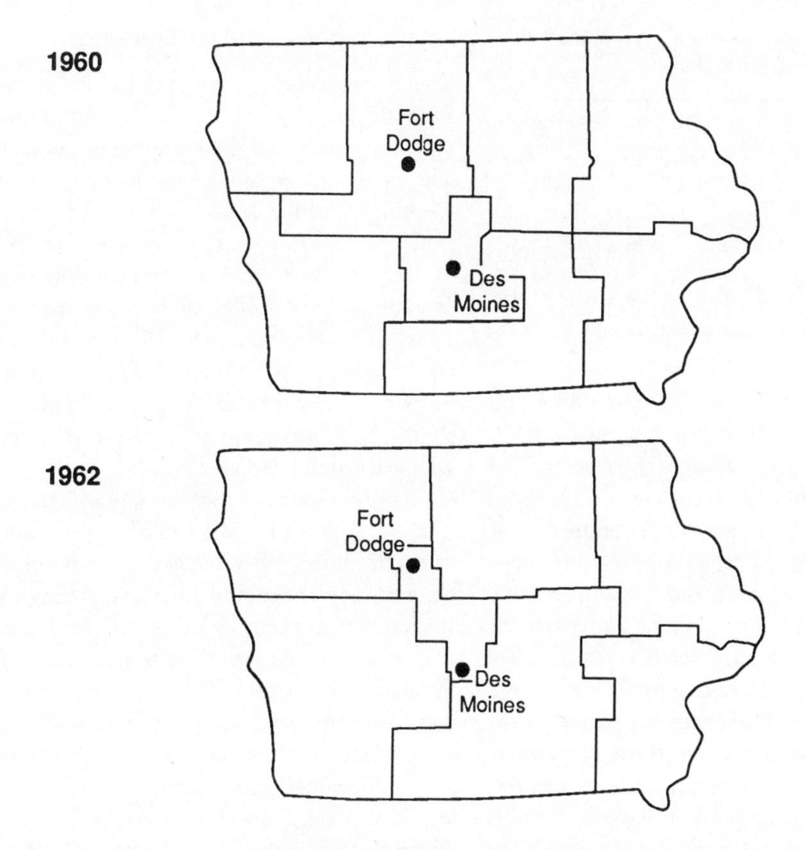

Figure 6.10 Gerrymandering in Iowa

system; that is, the candidate who gets the most votes in any constituency wins the election, becomes a member of parliament (MP) and the political party with the most MPs governs the country.

Table 6.4 shows the relationship between votes and seats in the 1987 general election. Note how the two main political parties, Labour and Conservative, gained proportionately more seats than votes. The biggest disparity occurred for the centrist Alliance which gained over a fifth of total votes but achieved less than twentieth of the seats. This disparity occurred because Alliance support was widely spread but rarely deep enough in any particular place to return an MP. The majority parties, in contrast, had support which was both wide and deep. As Table 6.5 shows it took less than 30 000 votes for a Conservative to win a seat, but an Alliance candidate had to pull in over a quarter of a million votes. The British example shows how electoral systems mediate between the intentions of voters and the political outcomes of elections. Such anomalies are unlikely to be changed by political parties who benefit from present arrangements. Governments can find all sorts of excuses why electoral reform is unnecessary and minority parties can find all kinds of reasons why they should be implemented.

The geography of electoral systems is very dependent on the type of electoral system. Three kinds of system can be identified: plural, proportional and preferential (see Table 6.6). The *plural system* is found in the UK and the USA. Let us look at some other systems. The *single transferable vote* system is one of multi-member constituencies and allows minority representation. It has been used in

Table 6.4 Distribution of votes and seats in the 1987 General Election

	Votes %	*Seats* %
Conservative	42.3	57.6
Labour	30.8	35.2
Alliance	22.6	3.3
Other	4.3	3.9

Eire since 1921 and takes the following form – one seat per 20000–30000 population with each constituency having a minimum of three seats. The maximum is now five. Electoral abuse can occur even in this system. The party in government can use their power to influence the size of constituencies. Smaller constituencies mean less seats, hence minority parties need more votes. The party in power will seek to draw up small constituencies where they have concentrated support and larger constituencies in areas where their support is weak.

In West Germany there is a form of *proportional representation.* There are single member constituencies and then a pool of seats distributed on the basis of overall votes cast by province. It is sometimes referred to as the *mixed system* or *additional member system* (AMS). It was one reason behind the electoral success of the Green Party throughout the 1970s and 1980s. The Greens were a minority in most constituencies and under a first-past-the-post system would have been victims of electoral democracy. However, because of the AMS system the Greens picked up seats, achieved electoral success, political instability and credibility.

Table 6.5 Votes per seat in the 1987 General Election

	Votes per seat
Conservative	29333
Labour	34934
Alliance	263360

Geography of spending

The government of a country has many functions to perform. One of the most important is the raising of taxes and the spending of public money. The state is a huge revenue-generating and spending machine. In this section I want to concentrate our attention on the spending side. The sums are impressive. Governments have huge amounts of public money at their disposal. In the fiscal year 1990–1, for example, the UK government spent £179 billion, approximately £3200 per year for every man, woman and child in the land. Spending by the USA government is in the trillions, sums which are almost beyond belief.

The spending of public money has definite spatial implications. We can identify a number of different types. A distinction can be drawn between *variate* and *invariate* spending. Variate spending means that by its very nature some forms of public spending are distributed unevenly over the national territory. Take the case of a government forestry programme or a soil conservation programme. Both these programmes are targeted at specific parts of the country because the projects are locationally specific. Invariate spending is where there is supposed to be no location specific project. I write 'supposed' because in reality other influences come into play.

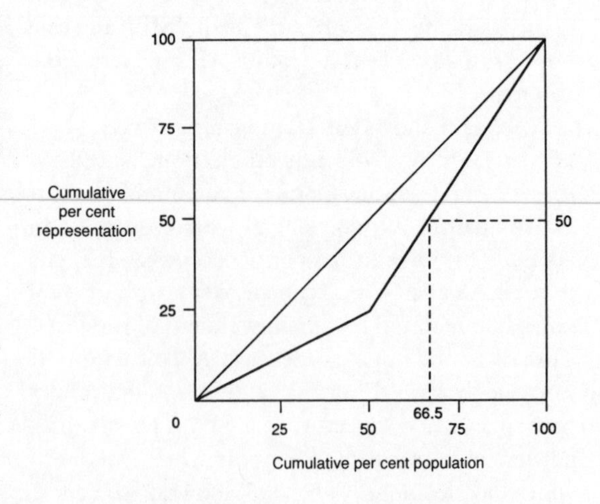

Figure N.1 Cumulative frequency graph

BOX N: MEASURING MALAPPORTIONMENT

Johnston and Taylor (1979) identify four methods of measuring malapportionment:

1. Compare the average size of constituencies won by different parties. In the case of Figure 6.9b, for example, the large, western *Red* area has 100 voters while the three eastern *Blue* constituencies have 33 or 34 voters. The difference, 77 voters, is a measure of malapportionment.
2. Measure the ratio of largest to smallest constituency. In Figure 6.9b this is 100/33 = 3.3. In other words the votes in the three eastern areas are worth 3.3 times more than votes in the large western area. In a system with no malapportionment the ratio would be close to zero. The larger the ratio the greater the malapportionment (see Table 6.2).
3. Compare every constituency with the average size to find the mean deviation. With no malapportionment all constituencies are the same size and the mean deviation is zero. In the case of Figure 6.9b the average size is 50, the total deviations are 50 + 17 + 17 + 16 = 100; the mean is thus (100/4 =) 25.
4. Construct a cumulative frequency graph. Figure N1 shows the results from the data in Figure 6.9b. If representation was equal to population the result would be a straight line diagonal. The graph to the right of this diagonal is known as a *Lorenz curve.* It gives us various measures of malapportionment. The area between the curve and the diagonal, for example, is a gross measure of malapportionment. When expressed as a proportion of the total area under the diagonal it is known as the *Gini index.* The *minimal majority* measures the smallest proportion of the population that can elect a majority of representation. It is found by drawing a line along the 50 per cent representatives column as shown in the figure, noting where it intersects the Lorenz curve and reading off the subsequent cumulative population. The minimal majority is then found by taking this figure from 100. In our example the figure read off from the graph is 66.5 hence the minimal majority is (100 − 66.5 =) 33.50. In other words, just over a third of the population could elect a representative majority. Table N1 shows the changes in the value of minimal majority in the US reapportionment revolution.

Table N.1 Changes in representation of state legislation, 1962–8

	Minimum % of population that could elect a majority to the Senate	
	1962	*21968*
Alabama	25	48
Arizona	13	52
Idaho	17	47
Maryland	14	47
Nevada	8	50
Rhode Island	18	50

Even programmes which are pursued for national ends have uneven effects. Take the case of defence, which is a national policy designed to safeguard all the territory of the nation-state. We may quibble about the term defence since defence policies invariably involve the arming of soldiers and the production of weapons of death and destruction. For the moment let us simply note this linguistic paradox and concentrate on the spatial consequences of defence. There is the obvious location of defence establishments – naval bases, army camps, etc. What is less obvious is the extent to which defence procurement tends to favour

Table 6.6 Three kinds of system

Plural	*Preferential*	*Proportional*
Candidates with most votes win, whether or not they have a majority of the vote.	Voter has chance to express second choice either in the same or another ballot, if no candidates have absolute majority on the first count.	Seats shared by parties according to their share of the vote.
Examples	*Examples*	*Examples*
Single member first-past-the-post e.g. USA, UK.	Alternative vote (AV) e.g. Australia; second ballot e.g. France.	Single transferable vote (STV) e.g. Ireland; list systems e.g. most of EC.
Advantages	*Advantages*	*Advantages*
Simple; short ballot paper; single-member seats possible; usually ensures one party has parliamentary majority.	Relatively simple; short ballot paper; single-member seats possible; candidates can stand without fear of splitting vote for right or left and so letting in minority opponent.	Fewer wasted votes; composition of legislature reflects voting figures; minorities represented. STV gives most choice of MP to voters and so loosens party grip on elections; list system simpler.
Disadvantages	*Disadvantages*	*Disadvantages*
Minorities under-represented (unless geographically concentrated); composition of legislature may not reflect national voting figures; party may have parliamentary majority based on minority of the vote. When used for multi-member seats, deficiencies exacerbated if party grip is strong.	Chance whether results more or less disproportionate than plural system: minorities still squeezed; AV requires voting by numbers; second ballot, two trips to the polls, and final result may depend on party bargaining.	Results depend somewhat on different formulae for allotting seats; constituences have to be large and so do ballot papers (although PR can be grafted on to single-member plural system, as in former West Germany). One-party majority in parliament unlikely. Requires long ballot paper and voting by numbers; list systems mean large constituencies and strong party control.

some regions rather than others. In Britain, for example, empirical work has shown that defence spending is heavily biased in favour of the south-east part of the country. The richest part of the country receives the lion's share of the £3 000 million spent on defence (Breheny, 1988). This government spending has kept the regional economy buoyant even during business downturns.

We are used to the terms sunbelt and rustbelt, but there are also gunbelt areas where local economies are biased toward hardware production for the military. Gunbelt areas are likely to experience a downturn as the cold war comes to an end.

Once a government spending programme is established it tends to generate a life of its own, continuing to generate the need for more funds. Since the bulk of government spending decisions are incremental, small additions to existing schemes rather than new projects, there is a built-in inertia factor. Most government spending goes to existing projects, which means there is a built-in bias to regions or locations already benefiting from government spending.

Governments target their spending for political purposes. We can make a distinction between the state, which is the continuing apparatus of power, and government, which is the political representation of those who yield power. Governments have many goals and objectives but one of their most important is to remain in power. Public spending is a powerful lever in this objective. Two different types of government influence can be noted:

- aiding supporters
- vote-buying

Let us look at these two in some detail.

Aiding supporters Given the opportunity all governments will target spending so as to aid their supporters. Consider central government support for local authority spending in Britain. Central government grants are allocated according to a complex formula of relevant statistics but the final equation is susceptible to political influence. When the Conservatives rule in Westminster the formula is fixed so that the gains are made by the rural and suburban districts which tend to be pro-Conservative. When Labour is in power in central government the grants formula is altered so that the urban authorities, which tend to be more pro-Labour, gain most.

Individual representatives also seek to use their power over spending decisions in order to reward their support. Elected representatives use their influence to direct public expenditure to their own constituencies. This is known as *pork-barrel politics* and it occurs throughout the world at all levels of government to differing extents. In Britain the exercise of pork-barrel politics is limited by the strong party structure and organization which restrict the actions of individual representatives. It is the politics of class which dominate the political scene. In the USA, by contrast, the politics of place take on a much more important role.

Pork-barrel politics are particularly important in the USA because the federal expenditure of an agency or a programme which produces a budget must be scrutinized by the relevant single and joint committees of the House of Representatives and the Senate. Committee members represent districts in particular states. The committees have a powerful role in determining the character and location of public expenditure. Members of the committee attempt to guide the flow of benefits toward their home states. Committee chairpersons are particularly influential. For example, the House of Representatives Armed Services Committee determines the location and size of military establishments in the different states of the Union. Up to 1970 the chairman of the Armed Services Committee represented a South Carolina district. By 1970, 0.67 per cent of the Defense Department's outlays went to this district, which had only 0.22 per cent of the country's population. To some extent the relationship between committee membership and the flow of expenditure is to be expected because representatives and senators will seek to get on those committees which deal with issues pertinent to the districts and the states they represent. However, Johnston (1978) studied the activities of the committees associated with NASA and the Atomic Energy Commission, which are not place-specific and for which there was no spatial pattern of need or resources. A study of these committees thus allowed the individual effects of committee membership to be measured. Johnston's results show that almost half of the state variation in expenditure was explained by the pattern of committee membership.

Vote buying Government spending is very sensitive to the geography of support. Governments do not want to spend too much money in areas where they have little support. For example, Conservative governments in Britain have little support in the inner cities and this fact explains the relatively low amounts of money directed to these areas of the country. However, governments are sensitive to marginal areas where spending may influence voting. Vote-buying is an old practice of governments and representatives. Let me illustrate with reference to Britain. In the early 1970s the Scottish Nationalist Party was gaining support in central Scotland. Popular feeling held that the Labour Government, which had a tiny parliamentary majority, was doing too little to aid the Scottish economy. In 1975 the government spent £150 million to bail out the Chrysler car company's plant in Scotland. It was a political not an economic decision.

Another example: in 1977 a by-election was to be held in the marginal constituency of Grimsby on 28 April. On 14 April three areas were raised to Development Area status, involving increased government aid to industry. Grimsby was one of the areas: Labour won the by-election.

And another: at both the local and national level Conservative policy during the late 1970s and 1980s was to sell off council houses to sitting tenants way below market levels. In effect it was a direct subsidy to council tenants who traditionally tended to support Labour. This windfall gain persuaded many tenants to support the Conservatives. In one London borough, Westminster, the policy was targeted at specific wards which were considered marginal. The policy was successful in turning many previous Labour supporters into Tory (Conservative) voters.

In conclusion, then, spending power is used to maintain and attract electoral support. Governments have huge amounts of public money at their disposal. The largest proportion is committed to existing projects and most of it is spent by civil servants and permanent committees. There is, however, a relatively small, though absolutely large, amount which can be directed by the party in power. Such money is sometimes directed by political parties to maintain allegiance in traditional areas of support and to swing the balance of elections in marginal constituencies.

GUIDE TO FURTHER READING

The section on people and environment draws heavily upon:

Short, J. R. (1991) *Imagined Country: Society, Culture and Environment.* Routledge, London.

Boundaries and frontiers are examined in various books:

Blake, G. and Schofield, R. (eds) (1988) *Boundaries and State Territory in the Middle East and North Africa.* Menus Press, Wisbech.

Eyre, R. (1990) *Frontiers.* BBC Books, London.

Prescott, J. R. V. (1985) *The Maritime Political Boundaries of the World.* Methuen, London.

Prescott, J. R. V. (1987) *Political Frontiers and Boundaries.* Allen & Unwin, London.

Rumley, D. and Minghi, J. V. (eds) (1991) *The Geography of Border Landscapes.* Routledge, London.

The geography of elections is a fascinating topic. Amongst the many books have a particular look at:

Johnston, R. J. (1987) *Money and Votes.* Routledge, London.

Johnston, R. J. (1979) *Political, Electoral and Spatial Systems.* Clarendon Press, Oxford.

Johnston, R. J., Shelley, M. and Taylor, P. J. (eds) (1990) *Developments in Electoral Geography.* Routledge & Kegan Paul, London.

Leonard, D. and Natkiel, R. (1987) *The Economist World Atlas of Elections.* Hodder & Stoughton, London.

Taylor, P. J. and Johnston, R. J. (1979) *Geography of Elections.* Penguin, Harmondsworth.

An analysis of public spending is available in:

Bennett, R. J. (1983) *The Geography of Public Finance* Methuen, London.

Breheny, M. J. (ed) (1988) *Defence Expenditure and Regional Development.* Mansell, London.

Heclo, H. and Wildavsky, A. (1981) *The Private Government of Public Money.* Macmillan, London.

Johnston, R. J. (1980) *The Geography of Federal Spending in the United States of America.* Wiley, Chichester.

Kodras, J. E. and Jones III, J. P. (eds) (1990) *Geographic Dimensions of US Social Policy.* Edward Arnold, London.

Wildavsky, A. (1986) *Budgeting: A Comparative Theory of Budgetary Processes.* Transaction Books, New Brunswick.

Relevant journals

Electoral Studies
Political Geography Quarterly

Other works cited in this chapter

Baker, G. E. (1966) *The Reapportionment Revolution.* Random House, New York.

Butler, D. E. and Stokes, D. E. (1969) *Political Change in Britain: Forces Shaping Electoral Choice.* Macmillan, London.

Grundy-Warr, C. and Schofield, R. N. (1990) 'Man-made lines that divide the world', *Geographical Magazine,* LXII, 10–15.

Johnston, R. J. (1978) 'Congressional committees and the geography of federal spending in the USA: the examples of NASA and AEC', *Area,* 10, 272–8.

Kolodny, A. (1975) *The Lay of The Land.* University of North Carolina Press, Chapel Hill.

Shoard, M. (1987) *This Land is Our Land.* Paladin Grafton, London.

PART III

THE POLITICAL GEOGRAPHY OF PARTICIPATION

So far we have discussed political geography almost entirely in relation to the evolution of the world order and the functioning of the nation-state. This is very much of a top-down view of the world; it is the view from the executive offices, the world as seen from the top floor of powerful corporations and the upper levels of government hierarchies. It says very little about the role of ordinary people. Political geographers rarely pay attention to the political geography of everyday life. And yet, there is a whole range of interesting questions concerning the relationship between the public and the private spheres, about the connections between social action and public policy, which are ignored if we confined our attention to the top-down functioning of the world order or of the state.

In this section I consider a more bottom-up view of the world order, the nation-state and the local state; a political geography which considers the active participation of people.

7

PEOPLE AND THE STATE

> People crushed by laws have no hope but from power. If laws are their enemies, they will be enemies to laws; and those who have no hope and nothing to lose will always be dangerous.
>
> (Letter to C. J. Fox, 8 October 1777)

TYPES OF STATE

In chapter 6 we considered the relationship between people and the state. It was a relationship of confirmation, a stable, essentially harmonious interaction between the governed and the government, in which citizens fulfilled their electoral responsibilities as laid down by the state. The relationship between the state and the population, let us call it civil society, is more complex and more interesting than this. The relationship can be identified as a continuum from outright acceptance on one extreme to outright rejection on the other; from passive acceptance of the status quo to outright conflict. Similarly a continuum of governmental responses can be identified from participatory democracy to authoritarian repression.

Table 7.1 Levels of stability

Type of state	*Examples*
Very stable	Netherlands, Switzerland
Stable	USA, UK, Canada
Unstable	El Salvador (1979–?)

A whole variety of state–civil society relationships thus exists; we can identify three broad types of state (see Table 7.1). In *very stable* (e.g. Netherlands, Switzerland) states there are few crises and there is an essential harmony between the state and the population. *Stable* states (e.g. Canada, the USA and the UK) are not racked by continual tensions between the government and the governed but conflicts can occur. Every ten years in Canada there seems to be a major constitutional crisis as the separatist movement in Quebec waxes and wanes. In both the USA and the UK urban riots have punctuated the political scene and in Northern Ireland the British state has a constant crisis of political legitimization. In unstable states crisis is the norm rather than the exception. In the case of El Salvador, for example, there has been civil war since 1979. The regime backed by the US government is very unpopular and repressive. From 1979 to 1990 nearly three-quarters of a million died and one million were displaced or exiled. Out of a total population of five million that represents an endemic crisis in political relationships. In unstable states the normal rules of political negotiation are suspended or denied, there may be competing centres of political power and ultimately social breakdown may erode the very fabric of civilized life.

These categories are not permanent. Things change. In the eighteenth century North America was in a state of political crisis as some colonists fought against British political control. In the nineteenth century the British state was in continuous struggle against the emerging working class. In both the USA and the UK, then, instability has been present. Unstable states may become more stable and stable states may become unstable if the crises outlined in chapter 4 become too severe.

PEOPLE AND STATE POWER

The relationship between the people and the state varies through time and across space and may involve different sections of the population in different ways.

At any one time across the surface of the nation-state there is likely to be a range of attitudes that the citizenry have toward the state. In very stable states, for example, most citizens will accept the legitimacy of the state most of the time. This will also be the case in stable states but conflicts may occur. They may be short term, localized affairs involving particular sections of the population in specific places. Such conflicts may arise, for example, over resistance to a new airport or complaints over the pollution levels from a nearby chemical factory. This conflict is focused on a particular issue (see chapter 9). There is also a resistance to central authority which may be deeper and longer. As we have already noted in chapter 5 most states are multi-ethnic, multi-regional affairs; when there is increasing incongruence between the separate nations of a state the rule of the state may not be accepted in all the different nations. This lack of legitimization may be seen in such symbolic acts of resistance as in Scotland where the 'British' national authority does not have the same capacity to evoke loyalty as it does in England and at international football matches many Scottish fans jeer the 'national' anthem.

More extreme responses can also be found. One example is the *intifada* in Israel. In the Arab–Israeli war of 1967 Israeli troops seized the West Bank from Jordan, the Gaza Strip from Egypt and the Golan Heights from Syria (see chapter 6). In the first two areas there were over one million Palestinians, many of whom had fled from Israel between 1948 and 1950. Dispossessed of their land, deprived of citizenship and ruled by what they saw as an alien, oppressive state the Palestinians of the Gaza Strip and the West Bank never accepted the legitimacy of Israeli rule. Tension was heightened in the West Bank with the establishment of over 140 Jewish settlements, which gave homes to almost 100 000 Jewish settlers on Arab land.

The situation was a tinder box waiting for a spark. On 9 December 1987 some Arabs were killed in a traffic accident at the Erez road block to the Gaza Strip. This became the rallying point for a spontaneous process of Arab resistance to Israeli rule throughout the occupied territories. The *intifada*, or uprising, involved a boycott of trade, strikes and civil disobedience, including the stoning of troops and police and attacks on the cars of Jewish settlers as they passed through Arab areas. The *intifada* has proved costly. On the one side, almost 400 Palestinians have been killed. On the other, Israel has lost public face as world-wide TV news pictures showed Israeli troops firing at school children or breaking the arms of stone throwers. The *intifada* has shown that popular resistance can influence both national and even international politics.

Let us look at popular resistance to state authority in some more detail. We can identify three important and related aspects:

- the context of protest
- the making of protest
- the consequences of protest

The context of protest

The state has much power and great influence. It is not an enemy to be taken on lightly. People need a very good reason to resist its authority. Let us look at the major reasons, with specific examples:

They may not accept the legitimacy of the state. For this basic conflict to be maintained there needs to be a major cleavage between an alienated population and the state. In the case of the *intifada*, religious and ethnic differences were overlain with the facts of recent history and the conflict over land and political rights. Other examples include the Catholic community in Northern Ireland, the majority of whom do not accept the legitimacy of British rule, and a tiny minority of whom resort to acts of violence and murder.

They may accept the basic legitimacy of the state but want someone else in charge. As a member of parliament in England said in 1693, 'It was not against centralized power that we fought but in whose hands it was.' A more recent example: in 1965 Ferdinand Marcos became president of the Philippines, and was re-elected in 1969. An astute politician, he was a fervent anti-communist and an important ally of the USA. The country contained US military bases and President Johnson even described him as 'our strong right arm man in Asia'. He was also one of the tackiest dictators around. After his re-election he used his power to rip the country off on a grand scale. Between 1972 and 1986 he siphoned off $20 billion, almost half of the country's entire gross national product. His wife, Imelda had almost three thousand pairs of shoes, and each pair cost the equal of the average Filipino's *annual* income.

Popular resistance grew. In response Marcos suspended the normal political process by declaring martial law. Radicals flocked to the New People's Army. This gave Marcos even more power. He persuaded successive US Presidents, including Nixon, Carter and Reagan, that this was communist insurgency. Through the prism of the cold war the USA aided Marcos, providing arms and aid. But Marcos was not overthrown by a communist-inspired plot. Nor was he deposed by a US government embarrassed by supporting such a figure. He was overthrown by ordinary Filipinos. When he tried to rig the result of the 1986 election they took to the streets in their hundreds of thousands and made his position untenable. The success of the people's politics was a result of many factors. Marcos was loathed. Even a staunch Republican like P. J. O'Rourke could describe the situation like this:

> Reporters who do duty in the third world spend a lot of time saying, 'It's not that simple'. We say: 'It's not that simple about the contras and the Sandinistas'. But in the Philippines it was that simple. It was simpler than that. Ferdinand Marcos is human sewage, an evil old power-addled flaming Glad Bag, a vicious lying dirtball who ought to have been dragged through the streets of Manila with his ears nailed to a dump truck.
>
> (O'Rourke, 1987, 74)

The opposition candidate in the election, Cory Aquino, had a large measure of support. The attitude of the USA was also important. The CIA thought a more effective counter-insurgency campaign against the People's Army could now be waged without Marcos. Just before the 1986 election the CIA released information which sought to show that Marcos was not the war hero he claimed to be. The President, in contrast, remained loyal to Marcos. Reagan continued to praise him, even as he had to flee the palace, leaving behind the gold-plated jacuzzi, the gold-leaf furniture and all those pairs of shoes. The US plane which took him and his family to Hawaii took assets of over $10 million and 48 feet of pearls. In the Philippines the people had been successful in achieving the overthrow of a dictator and replacing him with what looked like a more democratic government.

They may accept the legitimacy of the state but want a major change in direction. This desire for change can occur because of:

- *relative immiseration*: people feel themselves to be getting materially worse off
- *denial of rights*: which can involve both economic and political rights. Let us consider the case of the civil rights movement in the USA.

Racial segregation was a common practice throughout the South in buses, restaurants, public education, housing and even access to political rights. America was segregated by custom and by law. It was also reinforced by violence. On 28 August 1955 a young boy, Emmet Till was killed by two men in Mississippi. He was black and they were white. His 'crime'? He 'talked fresh' to a white woman. He said 'Bye baby' to her as he left a store. They came to his uncle's home where he was staying, beat him, mutilated him, then dumped his

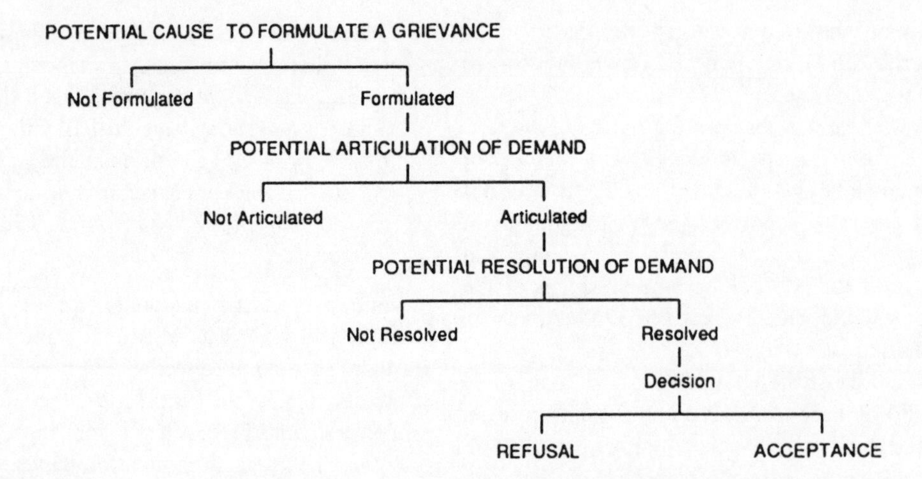

Figure 7.1 The process of political exclusion

body in a river. An all-white jury took only one hour to find the men not guilty.

Inequality was always resisted, but after the Second World War the resistance was more effectively organized. Black organizations such as the National Association for the Advancement of Coloured People (NAACP), first established in 1910, sought redress through the courts. In 1954 the Supreme Court ruled that school segregation was illegal.

On 1 December 1955 Mrs Rosa Parks left her work in Montgomery, Alabama. She boarded the bus to go home and sat down. The bus was crowded and, when more passengers got on, the driver asked Mrs Parks and three other people to move to the back of the bus. Mrs Parks and the three others were black and in Montgomery, Alabama in 1955 buses were segregated. When the bus driver asked Mrs Parks to move, she refused and was arrested: so began the bus boycott in Montgomery. Blacks, the main customers of the service, refused to use the buses. They walked or organized car pools. People were arrested and intimidated but the boycott held through most of 1956 until November when the bus company conceded the issue. Local bus segregation ordinances were repealed and a test before the Supreme Court prohibited segregation on all public buses.

The Montgomery bus boycott had a number of characteristics which were repeated throughout the USA during the 1950s and 1960s:

- acts of peaceful civil disobedience were undertaken by ordinary people. These many brave acts, often in the face of threats and violence, provided the backbone of the civil rights movement. Boycotts, marches and sit-ins throughout the country showed the depth of and commitment to social and political change.
- the movement was fortunate in having a charismatic leadership which gave direction to popular discontent and effectively lobbied the federal government. One of the leaders of the Montgomery bus boycott was the Reverend Martin Luther King who enunciated the philosophy of passive resistance with so much eloquence and passion.
- the civil rights movement at the local and state level was aided by the federal authorities. The 1954 Supreme Court ruling, for example, on school segregation was a potent sign to the black community. Advances were recorded by subsequent court rulings and legislation. The 1964 Civil Rights Act prohibited discrimination, while the 1965 Voting Rights Act gave blacks more political power, but the civil rights

legislation was ultimately the work of a concerned citizenry. It was the result of bottom-up pressure exerted by ordinary people doing brave things.

The making of protest

Ultimately, protest arises from the fact that a group of people have a grievance. However, a grievance may not necessarily result in protest, let alone mobilization. The expression of the grievance depends upon a number of factors (see Figure 7.1).

A group may have the basis for formulating a grievance because of, say, inadequate political representation or denial of rights, but the grievance may not be formulated if there is not a suitable organization to crystallize individual complaints into group action. For any social movement to be successful there needs to be a committed group of activists. The prevailing ideology may not even allow complaints to be perceived as group grievances. They may be seen as a function of individual failure. The affected groups themselves are often debilitated by the partial acceptance of these views. If the grievance is formulated, there is a possibility for the associated demands to be articulated. The demands may not be articulated because, for many alienated groups, there is the anticipated reaction, grounded in past experience, of 'what is the use, they never do anything for us'. It is an understandable position to take if you are at the bottom of the social hierarchy with no hope of improvement; the whole of your life experience reinforces a fatalistic attitude. Even if the demands are articulated, the issue may not be resolved by the political elite. Non-decision-making affects disadvantaged groups in two ways.

First, the groups have limited money and resources to press their claims. If the elite can avoid making quick decisions, either deliberately or because of unavoidable delays, then the pressure is likely to diminish and the articulated demands of lower-income groups will tend to fade from the immediate political scene. Second, no decision may be made because the elite may not perceive the articulated demands as legitimate claims. Stigmatization of housing action groups, for example, by putting the blame on 'outsiders', 'queue-jumpers' and 'political trouble-makers' is a common feature of politics. Finally, if the pressure group is able to get its demands resolved the decision may be to refuse its claim. It is at this stage that the expression of grievance may break out into action.

Table 7.2 A repertoire of collective action

Action	*Category*
Surveys, collection and presentation of evidence Petitions	Persuasive
Lobbying of local government officers, councillors, ministers, government departments and other decision-makers Mass letter-writing On-site discussion and demonstration	Collaborative
Fighting individual cases Deputations Rallies Marches Refusal to pay Civil disobedience	Confrontational

A repertoire of collective action can be identified. Table 7.2 shows the range. The actions range from:

- *persuasive*: where people seek to persuade state authorities of the need to do something
- *collaborative*: strategies adopted because the group has a shared set of basic assumptions with the authorities
- *confrontational*: where the group seeks to, or has to, confront the authorities.

The further away the social group involved is from influencing power-holders and the greater the grievance, the more likely it is that confrontational strategies will be used. People take to the streets when they have nothing to lose and lots to gain. Riots have been described as a festival of the oppressed. As the Reverend Martin Luther King once said, 'a riot is at bottom the language of the unheard'.

The consequences of protest

Social protest does not necessarily lead to social change. We can identify three types of consequence:

- protest unsuccessful
- protest partially successful
- protest successful

Protest unsuccessful. Throughout 1989 there was mounting student resentment in Beijing against the communist government. Economic reforms had generated tremendous inequalities and allowed corrupt bureaucrats to get rich. Students wanted social change and more democracy. In protest against the government they occupied Tiananmen Square in the centre of the capital. Throughout May and June a festival atmosphere began to develop in the giant square. On the night of Saturday, 3 June soldiers of the People's Liberation Army fired indiscriminately into the packed square. Estimates vary. The most conservative accounts say over 300 people were killed and almost 3000 were wounded. The power of the state had been used to quell the protest. The Avenue of Eternal Peace, which runs into the square, is now called by locals Blood Avenue.

There is a Sherlock Holmes story in which the great detective solves the case by noting the absence of something: the dog that did not bark. Political geography would be enriched by the study of the social equivalents of dogs that did not bark, for example:

- social contexts which should have, but did not, produce successful protest movements. The failure of genuinely democratic revolution in most of Central and Latin America would provide one such study.
- protests which were not successful. Why was there a revolution in Russia in 1917 but not one in 1905? Why did the left-wing revolution fail in Germany in 1917? The study of such failures allows us to understand the successes.

Protest partially successful. Few social movements achieve all their ends. Let us return to an earlier example. One of the most successful in recent years was the civil rights movement in the USA. The movement enabled blacks to obtain greater social and political rights. However, blacks in the USA still have lower incomes, poorer housing and higher infant mortality rates than whites. Economic rights have been harder to achieve.

Conversely, even unsuccessful protests often have their successes. The brutal repression of the students in Tiananmen Square swung world public opinion away from support for the ageing communist leadership. Events showed both the power of the state and the nervousness of the leaders, their resistance to change and their fear of free discussion. They were so afraid they felt they had to send in the troops. The party hierarchy gained control ... but for how long?

Protest successful. The study of public protest and collective action is an important corrective to those top-down studies which restrict their understanding of political change to the national workings of the state. Bottom-up changes have had a huge influence in changing particular policies, governments and even whole political systems. Let us end this chapter with a brief look at the really big changes.

France 1789, Russia 1917 and Eastern Europe 1989 are all examples of successful revolutions. They were revolutionary in the sense that there was a major rupture in state arrangements and civil society; in fact, a whole new social and political order was established.

Before 1789 France was a monarchy, people were subjects of the king; after the revolution they were citizens of a state. A social hierarchy with royalty at its apex was replaced by a society whose motto was 'liberty, equality and fraternity'. The changes were enormous, even in attitudes to time. From October 1793 a new calendar was introduced with twelve thirty-day months and five

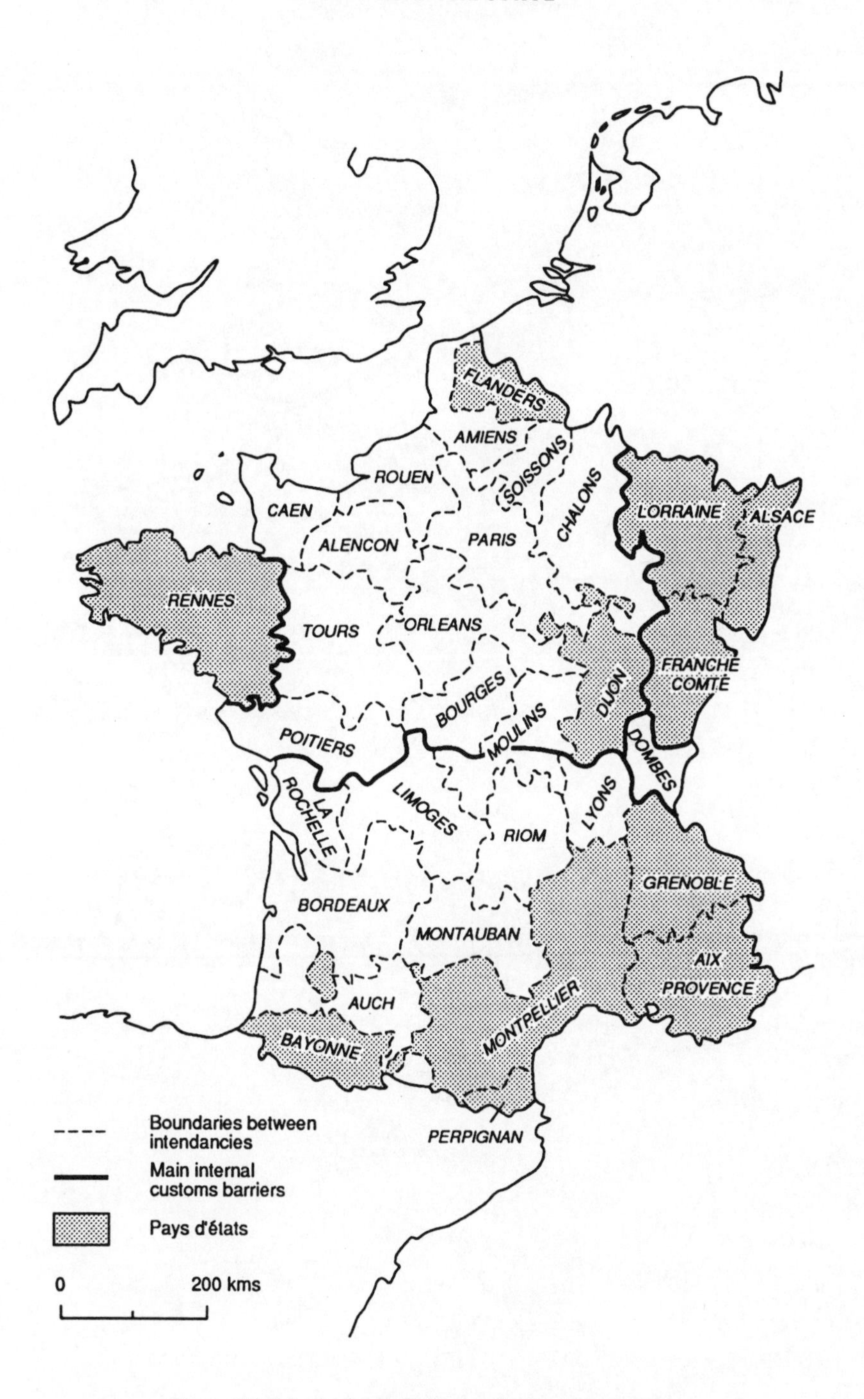

Figure 7.2 Principal sub-divisions of pre-revolutionary France

Figure 7.3 The departments of revolutionary France

Table 7.3 The French revolutionary calendar

Approximate time of year	*Revolutionary name*	*British name*
September/October	vendemiaire	slippy
October/November	brumaire	nippy
November/December	frimaire	drippy
December/January	nivose	freezy
January/February	pluviose	wheezy
February/March	ventose	sneezy
March/April	germinal	showery
April/May	floreal	flowery
May/June	prairial	bowery
June/July	messidor	heaty
July/August	thermidor	wheaty
August/September	fructidor	sweety

complementary days. Table 7.3 shows the new names, their equivalent in the Gregorian calendar and the names given by contemporary British commentators. The commentators thought they were being scornful, but I prefer the revolutionary names. Space was also transformed. Figure 7.2 shows the demarcation of France before the revolution. It was a country with internal customs boundaries and separate legislative systems. It was a space of separate places poorly integrated. Figure 7.3 shows the more integrated system introduced after the revolution. In effect a new state system of government was introduced, a whole new political geography was inaugurated.

In 1917 Russia was an empire ruled by a tsar, the economy was one of capitalism emerging from

BOX O: SOCIAL MOVEMENTS

A social movement is a group of people who come together for specific purposes because of a shared set of beliefs. Paul Wilkinson defines a social movement as:

1. a deliberate collective endeavour to promote change in any direction by any means
2. having a minimum degree of organization
3. with a commitment to change based on active participation of members

A whole variety of social movements can be identified, expressing religious sentiment, rural and urban discontent, nationalist and race movements, class movements, age and gender movements.

The most important aspects of any social movement are:

- *context*: how, why, when and where do movements occur?
- *organization*: how is the movement organized? Who runs it?
- *mobilization*: how does it gain resources and how does it exercise its power? What is the repertoire of available collective action?
- *opportunity*: where, when and how does it exercise its power?
- *consequences*: what are the results and effects of its action?

References

Tilly, C. (1978) *From Mobilization To Revolution.* Addison Wesley, Reading, Mass.
Touraine, A. (1981) *The Voice and The Eye.* Cambridge University Press, Cambridge.
Wilkinson, P. (1971) *Social Movement.* Macmillan, London.

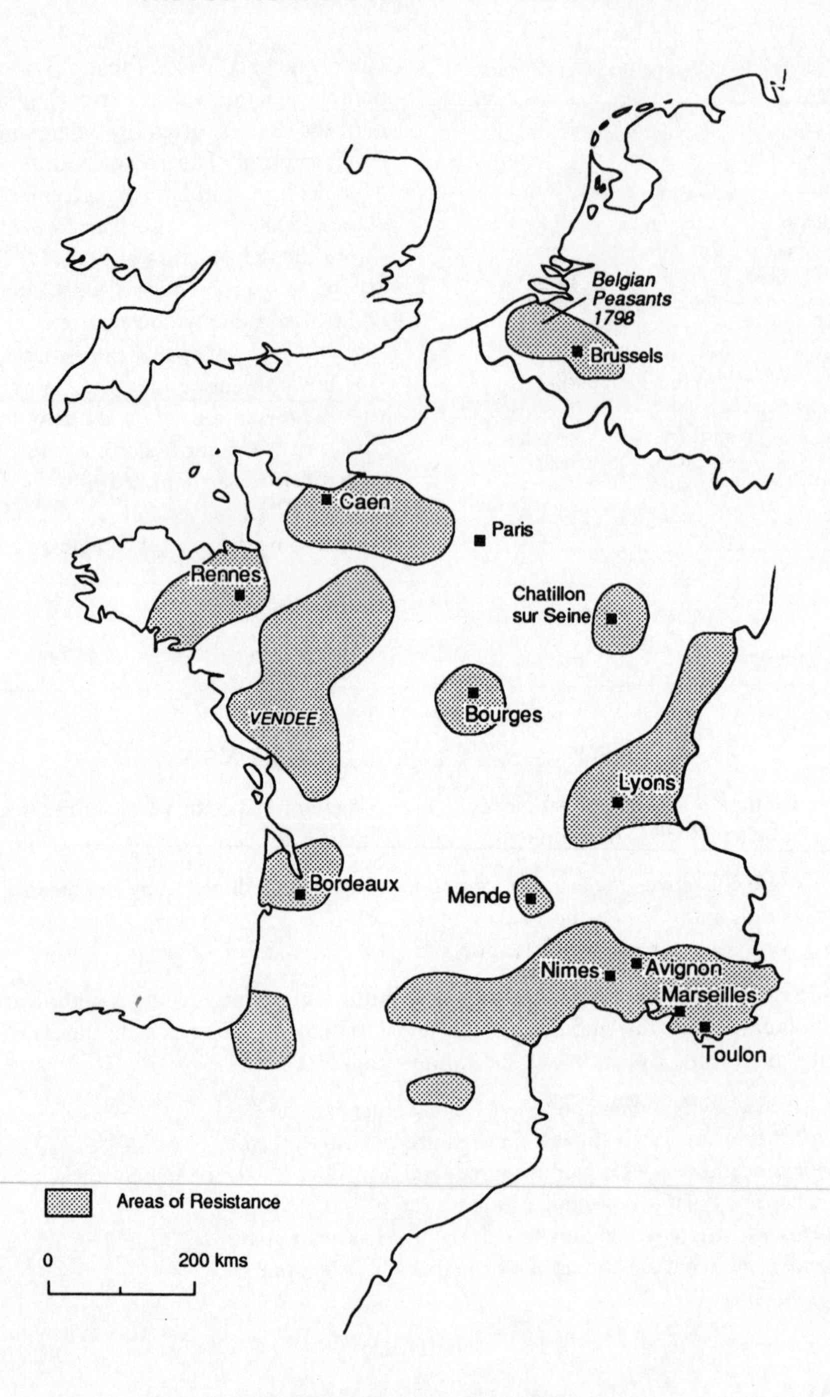

Figure 7.4 Resistance to the revolution, 1793–9

feudalism and politics was dominated by the royal household and the large landowners. After the Bolshevik revolution the existing social hierarchy was abolished, socialist central planning was introduced and the dictatorship of the proletariat was introduced. Subjects had become comrades.

Before 1989 Eastern Europe was ruled by communist party governments whose power was underwritten by Soviet military power. After 1989 comrades became citizens in multi-party electoral systems and market signals began to replace planning directions. Capitalism replaced communism and democracy began to replace dictatorship.

All were major changes. And all were changes brought about by *people power* involving the successful entry of mass movements into political discourse.

What makes people take to the streets? The first thing that must be remembered is that not everyone takes to the streets. After successful revolutions more people will say they took to the streets than actually did. Before success is assured protest is a potentially dangerous thing, limited to the very motivated, the very poor and the very desperate. What provides the motor of popular protest is popular discontent. As George Rude noted with reference to the French Revolution:

> revolutionary crowds cannot be dismissed as mere passive instruments of middle-class leaders and interests, still less can they be presented as inchoate 'mobs' without any social identity or, at best, drawn from criminal elements or the dregs of the city population ... far from being social abstraction (they) were composed of ordinary men and women with varying social needs who responded to a variety of impulses in which economic crisis, political upheaval, and the urge to satisfy immediate and particular grievances all played their part.
>
> (Rude, 1959, 232–3)

In late eighteenth-century France and in Tsarist Russia there was hostility towards modernization, an emerging capitalism was causing immiseration for the many. In Eastern Europe the queues were getting longer and the goods shoddier and ever more scarce. A sense of bitter frustration pervaded all three situations, a sense of alienation from the established order. And in all these cases protest became 'successful' because there was also a crisis amongst the governing elite, the rupture was partly a result of a split amongst those in power. Popular discontent on the one hand and political incompetence on the other do not always bring success but they are the essential ingredients.

There is a political geography to revolutions. Revolutionary sentiment tends to be stronger in some places than in others, and not everywhere shows the same degree of commitment to social change. Figure 7.4, for example, shows the centres of resistance against the French Revolution from 1793 to 1799.

Protest activity is concentrated in specific places. In many cases the big towns and cities have a culture of resistance which can be mobilized into popular protest. Even the expression '*taking to the streets*' captures the urban bias of social protest. In revolutionary France there was the Great Fear of the summer of 1789 in which much of the countryside was in revolt but revolutionary events took their sharpest turn and had their most dramatic effects in Paris. Similarly, in Eastern Europe it was the cities and particularly the capital cities, where protests against the communist governments were strongest, most dramatic and most successful.

Particular places become of tremendous importance in social upheavals. The storming of the Winter Palace, the taking of the Bastille, and the toppling of the Berlin Wall are all examples of symbolic events which both signified and codified revolutionary change. Significant events occur in significant places. Successful revolutions capture significant places and give them new meaning, a whole new symbolism. Space and place are not simply the passive background to politics, they are central to the exercise of political power, and the struggle for political power.

GUIDE TO FURTHER READING

On the relationship between people and the state have a look at some of the following:

Bright, C. and Harding, S. (eds) (1984) *Statemaking and Social Movement.* University of Michigan Press, Ann Arbor.

Foss, D. A. and Larkin, R. (1986) *Beyond Revolution: A New Theory of Social Movements.* Bergin and Garvey, South Hadley, Mass.

Krantz, F. (ed) (1988) *History From Below.* Basil Blackwell, Oxford.

Lofland, J. (1985) *Protest: Studies of Collective Behavior and Social Movements.* Transaction Books. New Brunswick, New Jersey.

Rude, G. (1964) *The Crowd in History.* Wiley, London.

Background reading for specific examples cited in the text include:

Israel

Said, E. and Hitchens, C. (eds) (1988) *Blaming The Victims.* Verso, London.

Schiff, Z. and Ya'ari, E. (1990) *Intifada.* Simon & Schuster, New York.

Shipler, D. K. (1986) *Arab and Jew: Wounded Spirits in a Promised Land.* Random House, New York.

Civil Rights Movement

Garrow, D. J. (1988) *Bearing The Cross.* Jonathan Cape, London.

Sitkoff, H. (1981) *The Struggle For Black Equality 1954–1980.* Hill & Wang, New York.

Williams, J. (1987) *Eyes On The Prize.* Viking, New York.

China

Fathers, M. and Higgins, M. (1989) *Tiananmen.* Independent, London.

On revolutions in general, sample from the following:

Goldstone, J. A. (ed) (1986) *Revolution: Theoretical, Comparative and Historical Studies.* Harcourt, Brace Jovanovich, Orlando, Florida.

Skocpal, T. (1979) *States and Social Revolution.* Cambridge University Press, Cambridge.

Wheatcroft, A. (1983) *The World Atlas of Revolution.* Hamish Hamilton, London.

For specific examples of the revolution in France, Russia and the USA consider:

Doyle, W. (1989) *The Oxford History of The French Revolution.* Clarendon Press, Oxford.

Middlekauf, R. (1982) *The Glorious Cause: The American Revolution, 1763–1789.* Oxford University Press, New York.

Rude, G. (1988) *The French Revolution.* Weidenfeld & Nicolson, London.

Schama, S. (1989) *Citizens.* A. A. Knopf, New York.

Trotsky, L. (1977, first published 1932–3) *The History of The Russian Revolution.* Pluto Press, London.

Wilson, E. (1940) *To the Finland Station.* Collins, London.

Other works cited in this chapter

O'Rourke, P. J. (1987) *Republican Party Reptile.* Pan, London.

Rude, G. (1959) *The Crowd in The French Revolution.* Oxford University Press, Oxford.

8

THE GLOBAL VILLAGERS

The world order we discussed in Part I was of nation-states and international economic systems. The main actors were multinational corporations and national governments. In this long-term perspective the actions of people, whether as social groups or individuals, were not really considered.

In the past a case could be made for a separation between a global focus on institutions and a more *local* emphasis on people and their actions. However, recent years have seen the development of a popular global awareness which collapses the distinction between the global and the local. How did it come about?

THE *GLOBAL* VILLAGE

Our view of the world is related to the means of communication. Our knowledge of the world is mediated by the technical means of information exchange. When we are limited by word of mouth our horizons are limited to the very local and distant events remain distant. With the development of print, messages can travel further and quicker and more reliably. Benedict Anderson (1983) considers the connections between the rise of print capitalism and the way nationalist sentiments are fostered. He uses the term '*imagined communities*' to refer to the way that a newspaper discourse can establish a community of interest.

With the development of electronic media the world is brought into our living rooms. At the flick of a button we can see events on the other side of the world *as they happen.* Indeed some events happen *because they are seen* on television screens around the world. We now live in what Marshall McLuhan referred to as the *global village.* We turn on the television and see riots in South Korea, revolutions in the Philippines, a general election in Germany, a disaster in Mexico, a war in Kuwait, an uprising in Iraq. With electronic media the world is brought into the perception of our everyday lives. In an age of electronic mass media knowledge of world events is no larger restricted to a select few.

There is now more of an immediacy to our experience of distant events. And there is a sharing of this experience. Radio and television has helped to create a shared world view and brought into play a major new force in world affairs – *world public opinion.* World public opinion can be mobilized to release political prisoners, undermine governments, legitimize opposition groups and give hope and help to beleaguered groups. For the first time in the world history the collective opinion of ordinary people has a role to play in global events.

This world opinion is as yet still very much restricted to the rich core. In North America there are two television sets to every household, in central Africa there is one set for every fifty households. The information presented on television is filtered and selective. Coverage of international stories is biased and patchy, subject to national stereotyping. Some groups are delegitimized by being called terrorists, others are lauded as liberation movements. Much of the Third World is only seen as newsworthy in the West if revolution, famine or natural disasters occur. Little attention is given to the everyday life of the majority of ordinary people. The foreign news of most national television news broadcasts concentrates on the highly unusual, the bizarre and the dramatic rather than the routine of mundane life. There are important differences in the way television news covers the local, the national and the international.

Despite this bias the last half of the twentieth century has seen a world brought much closer together, a world where more people now know much more about the rest of the planet, where world public opinion has emerged as a powerful force.

THE GLOBAL *VILLAGE*

The twentieth century also saw the realization that the world was a very small place and what happened in one part had an effect elsewhere. When Chernobyl exploded in April 1986, radioactive dust fell on the hillsides of North Wales; when people used aerosols in the privacy of their homes they are affecting the atmosphere above all our heads; and when people cut down trees in the Amazon, flood more rice fields or raise more cattle they affect the climate of the whole world. We now know that although we may be separated into different states these boundaries are becoming more and more irrelevant. We have become *global villagers*, a term which captures the interdependency of the global and the local. Our everyday lives are lived at a local level, our actions are, on most days, bounded by a fairly tight spatial spread. And yet we are connected to the rest of the world, connected by news and information and the consequences and implications of our actions. There is a connection between the global and the local.

Let us consider three particular areas of interest to the concerned global villagers:

- war and peace
- poverty and plenty
- ecological issues

War and peace

On 6 August 1945 an American aircraft dropped an atomic bomb on Hiroshima, in Japan, killing 78000 people. Arthur Koestler (1980) suggested that dates should now be affixed with the initials AH (After Hiroshima) to distinguish the old world from the nuclear world. Only one more bomb has been dropped in anger, on 9 August 1945 on the Japanese city of Nagasaki, but the threat of nuclear war has hung over the planet just as the mushroom cloud hung over the two devastated cities.

In the late 1940s and early 1950s the threat of the 'other' was used to legitimize the nuclear arsenals of both the USA and USSR. The nuclear capability of Britain and France was a remnant of imperial delusions. But the public was never happy about nuclear weapons and there was a widespread public revolution against their use. Nuclear weapons were feared because most of the public, unlike the military, could see that they threatened the existence of human civilization and the long term occupancy of the planet. There was no such thing as a limited war with nuclear weapons. Radiation fall-out would spread throughout the world.

The military and political response on either side of the Iron Curtain was to rely on the *deterrence theory* – we need to have them in order to stop their use. This argument can be used to maintain existing systems but fails to convince the educated global villagers of the need for new systems. We can see the development of an anti-nuclear weapon stance most clearly in western Europe in the early 1980s. There was tremendous public resistance in most European countries against the development of Cruise missiles. Western Europe did not have the anti-communist rhetoric of the Reagan administration, and there was a legitimate fear that Western Europe would be reduced to a pile of ashes in a superpower nuclear exchange. When a group of women set up a camp outside Greenham Common in Berkshire, one of the UK's designated centres for Cruise missile deployment, they also established a powerful symbol of resistance. Peace marches and anti-Cruise demonstrations throughout the early and mid 1980s created a powerful climate of opinion against the build-up of nuclear missiles. Nuclear weapons are still being made and they form an important part of the weaponry of an increasing number of countries, but up to now states have not been willing to use them. Popular global opinion has made the deployment of nuclear weapons a contentious issue.

Poverty and plenty

As we noted in chapter 1 the world divided into rich and poor, countries which suffer the affluenza of too much wealth and those that suffer great poverty. This state of affairs rarely infringes on our daily lives. There are poor areas and very poor people in rich countries but unless one goes to the Third World one rarely see the appalling degradation of mass, endemic poverty. Even a visit to a Third World country does not necessarily bring such an experience. Life in the Hilton, whether it be in Rio or Rome, New York or Manila is much the same. And even if you step outside the air-conditioned lobby the experience is a personal one, it does not provide the basis for collective action.

Public attention in rich countries can be focused by the mass media. This tends to happen when there are major famines and mass starvation. Harrowing pictures of tiny children, their bellies swollen by malnutrition, always seem to strike a sympathetic chord within the rich countries. Pictures of Ethiopian children stimulated such charitable endeavours as Band Aid, songs such as *Feed The World* were sung across the whole planet and a great deal of money was raised.

Sometimes the aid produced by emotional responses are a short-term palliative, sometimes they are not effective at all. What they do represent, however, is the deep-seated feeling that we have a responsibility for one another, that people in trouble should be helped, the hungry should be fed and the homeless should be sheltered. Global public opinion can be and is mobilized by the social tragedies of mass famine or natural disasters.

Ecological issues

On Sunday, 22 April 1990 Earth Day was observed throughout the world. Around the planet more than 200 million people took part in global celebrations which ranged from marches, speeches and demonstrations to parties and dances as global villagers showed their concern with ecological issues.

The growing awareness that we live on a shared planet and the steady realization of the scale of environmental pollution and damage has sensitized concerned global villagers to the need for a global ecological awareness. This awareness operates at three levels:

1. local concerns with particular issues affecting the immediate area
2. attempts at influencing national policies
3. concern with major environmental issues in selected parts of the world, e. g. the clearing of the tropical rainforest, the depletion of natural resources and the danger to dozens of species of birds, animals and sea creatures

For many people it is easier to be concerned with point 3 rather than 1 or 2. Events further away seem easier to understand, less marked by ambiguity and thus make it easier to take a stance. Who in suburban New England, for example, is going to fight *for* the logging of the Brazilian rainforest? Closer up and nearer to home the issues become more blurred, more susceptible to competing interpretation. Thus in the late 1980s the Conservative government in Britain could pontificate about the need for preserving the tropical rainforest but still justify a road building programme in London which cut through ancient woodlands.

The different levels of awareness also draw upon different repertoires of collective action. Local and national concerns generate lobbying tactics from people lobbying national politicians. Global concerns tend to be more difficult to mobilize.

There are few forums for the exercise of global opinion. We do not have votes in world elections in the same way we do at the national level. However, we still have some degree of power – as consumers we have market power and as participants in demonstrations we can seek to influence the climate of opinion. It is highly unlikely, for example, that Nelson Mandela would have been released without the unblinking gaze of world public opinion.

It is difficult to mobilize public opinion regarding the long-term future of the planet or for world peace. It has proved much easier when a specific focus could be found. If we look at the major campaigns of the 1980s, whose aims were to:

- ban CFCs
- release Nelson Mandela
- stop the deployment of Cruise missiles
- give food to Ethiopia
- raise money for Amnesty International

we see that they all combined global issues of war and peace, poverty and plenty, human rights and ecological issues with specific goals. They also all shared a particular characteristic – they tended to be as much celebratory as provocative in nature. Whether it was pop concerts or 'fun runs' they mobilized through celebrations. Unlike demonstrations aimed at specific targets there was an element of coming together not to intimidate someone else, but to feel the collective will, to experience the global connections.

We are only just beginning to know that we are global villagers. We are only just beginning to sense our global rights and global responsibilities, our global role and our global opportunities for social change.

GUIDE TO FURTHER READING

On the creation of the global village see:

McLuhan, M. (1962) *The Gutenberg Galaxy: The Making of Topographic Maw.* University of Toronto Press, Toronto.

McLuhan, M. (1964) *Understanding Media: The Extensions of Man.* McGraw-Hill, New York.

On selected global issues examine:

Arnold, D. (1988) *Famine.* Basil Blackwell, Oxford.

McCormack, J. (1989) *The Global Environmental Movement.* Belhaven Press, London.

Pepper, D. and Jenkins, A. (eds) (1985) *The Geography of Peace and War.* Basil Blackwell, Oxford.

Other works cited in the text

Anderson, B. (1983) *Imagined Communities: Reflections on the Origin and Spread of Nationalism.* Verso, London.

Koestler, A. (1980) *Bricks To Babel.* Hutchinson, London.

9

CITIZENS AND THE CITY

There is no more serious menace to our civilization than our rabble-ruled cities.

(Josiah Strong, 1885)

The city is a social invention whose size, location and internal organization is a function of the distribution of power. The city reflects the struggle for power and many of the major changes in the city's role and structure come about because of the mobilization of the citizenry. The shape and form of the city, like the state, is partly a function of bottom-up pressure. In this chapter we will focus our attention on the role of social movements in the city.

GENERAL COMMENTS

Citizens

The building blocks of urban social movements are the citizens. In the political arena of the city citizens play a number of roles:

- as *workers*: people are employed in the city. As public service workers, they may be directly employed by the city authorities and are thus implicitly involved in city finances.
- as *tax payers*: households pay for local public services. They want to maximize the benefits and minimize the costs. This can bring them into conflict with those state authorities who want to increase local taxes, or even public sector workers who want higher wages.
- as *users of services*: households also use a range of public goods and services such as education, roads, parks, etc. As users they want to maximize benefits. This may bring them into contact with the producers of these goods and services, the public service workers, the city authorities and central government.
- as *residents*: households are residents in a double sense. They are residents of particular places – neighbourhoods with a cultural identity. They are also residents of general urban space. We can imagine this space as a changing *externality* surface. Positive externalities are the benign effects of changes, such as an attractive park or an improved transportation link. Negative externalities could include a 12-lane highway constructed just outside your home. We can identify *public goods* which cause positive effects on house prices and local property values and *public bads* which reduce property values. Households, as residents, are concerned with the quality of their lives and their property values. The city is a constantly changing externality surface because the location of public goods and bads is constantly changing. Residents will seek to attract public goods to their local areas and repel public bads. In terms of the latter, residents have the *exit/voice choice.* They can move in order to escape the negative effects. Much of the white flight to the suburbs from US cities, for example, was an attempt to escape from what was perceived as deteriorating living conditions in the inner city. Alternatively, residents can stay put and either accept the changes or voice

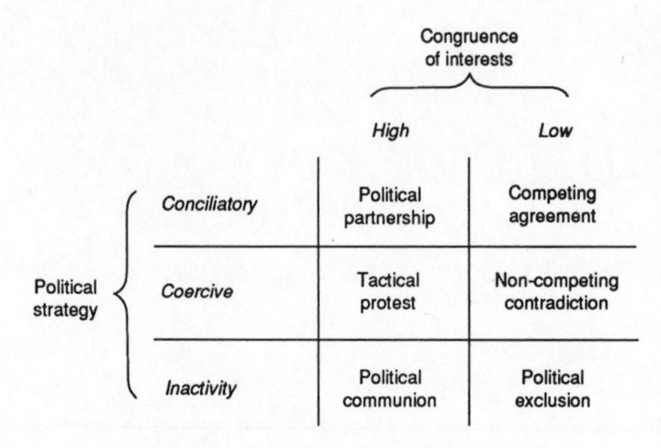

Figure 9.1 Typology of pressure groups

their disapproval, mobilize political support and influence political decisions. Their success depends upon their connections with the political elite, their organizational skills, the stakes involved and the strategies they adopt.

Pressure groups

The strategies adopted by urban social movements depend upon their relationship with the political elite. Peter Saunders (1979) has proposed a typology of movements based on their strategy and their connections (see Figure 9.1). His typology arises from two axes. The first refers to the congruence of interests, and measures the extent to which the demands of the group correspond to those of the political elite. The second axis refers to the strategies, and divides the pressure groups according to the methods they adopt to achieve their demands. The resulting typology provides a sixfold division. *Political partnership* occurs when the pressure group shares the same interests as the political elite and adopts a conciliatory attitude in achieving its demands. *Competing agreement* occurs when the pressure group adopts a conciliatory attitude but has only limited correspondence with the political elite. 'Respectable' community organizations wanting more resources for their area and going through the 'proper' channels would fall into this group. *Tactical protest* is the strategy of direct action employed by those groups who share similar interests with the political elite. In Britain, middle-class parents in Tory-controlled Croydon campaigning against comprehensive education and teachers' unions in Labour-controlled Sheffield campaigning *for* comprehensive education are examples from either side of the political fence. Often tactical protest is encouraged by the local political elite since it demonstrates to the media and the central authorities the strength of feeling behind their case. The ugly phrase, *non-competing contradiction*, describes the situation when the pressure group uses direct action but its goals are not shared by the political elite. Squatters, for example, and various action groups fighting against welfare cuts frequently encounter hostility from the political elite, who often classify them as 'irresponsible'. The designation is used to dismiss their claims and justify the stance of the elite.

Pressure groups are the explicit expression of demands from certain groups in society but there are some groups which do not need to or cannot express their demands because the political system is orientated towards respectively meeting and ignoring such demands. *Political communion* describes the state of affairs when a group shares the same interests as the political elite but does not need to become involved directly. Big business in particular does not need to exert pressure or rely on contacts because their interests are seen as the general interest. *Political exclusion*, in contrast, occurs when certain groups are not represented. This often happens to the very poor and the very weak.

The typology of pressure group types sketched out by Saunders with reference to Britain is broadly applicable to the situation in North America. In most urban municipalities there is a relationship of political communion between business interests and city councillors, and the political exclusion of lower-income groups is equally applicable. The debates within the USA, however, have largely focused upon showing which groups wield effective power in city politics. Three positions can be identified. The *elitist* position states that a self-conscious elite effectively runs the city; the exemplar work of Hunter (1953) pointed to the dominant position of businessmen in municipal government. The *pluralist* position states that power is diffused among a number of competing centres. No one interest or pressure group dominates all aspects of policy. This view is similar to the democratic-pluralist view of the state outlined in chapter 3. Dahl's work in New Haven purports to show that pressure group influence varies according to the policy issue (Dahl, 1961). It is interesting to note that Dahl's analysis of the three issues of urban redevelopment, public education and party nominations partly undermines his own case. To be sure, different individuals and pressure groups were involved in each of the three areas of public debate but the analysis did show the overwhelming importance of middle-income and business interests, albeit expressed by different individuals in separate organizations.

The *neo-elitist* position was formulated in response to the arguments of the pluralists. This position asserts that power does not lie at the level of pressure group activity but resides in the ability of certain groups to exclude a wide range of issues from public debate. Hayes's (1972) study of urban renewal in Oakland, California, for example, showed that urban renewal was promoted by business interests and the issues of the destruction of low-income housing and the displacement of the inner-city black population were not raised on the political agenda. Lukes (1974) has identified the main differences between the approaches (see Figure 9.2). He suggests that the pluralist approach takes a one-dimensional view of power with its emphasis on the behaviour of pressure groups and the conflict generated by particular issues. The neo-elitist position adopts a two-dimensional view of power since it focuses not only on decision-making and issues but also on non-decision-making and potential issues. What is needed, argues Lukes, is a three-dimensional view of power which can incorporate an understanding of observable and latent conflict, subjective and real interests, and issues and potential issues, since it is only such a wider view that can bare the relations of power within society.

One-dimensional view	Two-dimensional view	Three-dimensional view
behaviour	behaviour	behaviour
decision-making	decision-making and non-decision-making	decision-making, non-decision-making and control of political agenda
issues	issues and potential issues	issues and potential issues
observable conflict	observable conflict	observable and latent conflict
subjective interests	subjective interests	subjective and real interests

Figure 9.2 Perspectives on power

Lukes makes an important point but this three-dimensional view of power is not without problems. The main difficulty is encountered by the observer who seeks to discover latent conflict and real interests if these differ sharply from observed conflict and subjective interests. By definition they lie below the surface of the participants' perception; they cannot be grasped by an analysis of actual events but only from a privileged theoretical position. In this case there is always the danger of slipping into metaphysical speculation – actual events being measured and judged from a theoretical position impregnable to the attacks of empirical reality. This potential danger does not negate the use of the three-dimensional approach, as it is only the three-dimensional view which can achieve anything more than a superficial analysis of power relations, but it does suggest caution in its application.

As a useful failsafe Saunders (1979) suggests that the analysis should always consider the actual distribution of costs and benefits. Losers and winners can be identified even if they do not see themselves as losers or winners. Consider as an example the case of urban renewal in major US cities during the 1950s and 1960s. The winners were big business and construction companies. The losers were low-income, predominantly black households who faced further restrictions on their already slight housing opportunities as cheap inner-city housing was demolished to make way for commercial developments and more expensive housing. The benefits of urban-renewal policies accrued to finance and construction capital and the costs were borne by low-income households in the inner city. The Saunders suggestion is only a partial solution, since we have to define costs and benefits. This is not an easy task and one that demands an *a priori* theory of what constitute costs and benefits.

The major cleavages

Table 9.1 shows the major cleavages in contemporary urban movements. Let us look at these cleavages in some more detail by examining the goal of each movement.

The city as use value: many urban social movements are concerned with the city as a place to live, in contrast to those powerful groups who see the city as a place to make money. Conflicts can arise because some capitalists put profit before people. But conflicts are rarely fought in such large terms. More common is the struggle over specific issues, e.g. saving a park or an old building from a developer who wants to build a new, profitable office block. Struggles may also arise over such issues as the provision of public transport, the size and costs of public housing, the provision of recreational facilities.

Struggle for cultural identity: neighbourhoods are not only places to live, they are of tremendous significance for the cultural identity of specific groups. Indeed, the identity of certain groups may be intimately bound up with living in specific places. Ethnic areas, working-class districts, middle-class enclosures are all hard fragments of meaning. The attachments may be so great as to initiate a formidable defence if the area is threatened by redevelopment.

Citizen participation: citizens are involved in a power struggle. On the one hand local authorities and central government seek to control the meaning of cities, the rhythms of urban life and the size, location and density of urban infrastructure. On the other hand many citizens want greater

Table 9.1 The major cleavages

Goal of the urban movement	The city as use value	Identity, cultural autonomy and communication	Territorially based self-management
Ideological themes and historical demands included in this goal	Social wage Quality of life Conservation of history and nature	Neighbourhood life Ethnic cultures Historical traditions	Local autonomy Neighbourhood decentralization Citizen participation
Name of the adversary	Capital	Technocracy	State
Goal of the adversary	The city as exchange-value	Monopoly of messages and one-way information flows	Centralization of power, rationalization of bureaucracy Insulation of the apparatus
Conflicts over the historical meaning of city	City as a spatial support for life *versus* City as a commodity or a support of commodity production and circulation	City as a communication network and a source of cultural innovation *versus* Despatialization of programmed one-way information flows	City as a self-governing entity *versus* City as a subject of the central state at the service of world-wide empires

BOX P: AN URBAN RIOT

Just occasionally protest leads to a riot. So it was on 31 March 1990 in central London.

On most week-ends demonstrations take place in London. This one was a march from Kennington to Trafalgar Square to show disapproval of the Government's poll tax, a flat rate regressive tax which does not take into account different abilities to pay. Over 100 000 people took part in what appeared to be a very peaceful march. One man had a placard which read 'Normally law-abiding person against the poll tax' while another held up a banner with the words 'Elvis Fans Against the Poll Tax'. Trafalgar Square was mobbed with people. As one participant noted, 'The atmosphere in the square was almost carnival-like ... it was an effortless show of power. *We owned Trafalgar Square.*'

The trouble began round about 3 p.m. A large group stood outside Downing Street, the heavily-guarded home of the Prime Minister. Stones and bottles were thrown at the police, who sought to disperse the crowd. The trouble flared along Whitehall and into Trafalgar Square.

Pandemonium broke out, police were attacked, then police on horses charged. Some of the demonstrators fought back. Rocks were thrown, sticks were wielded. Injuries were caused.

Some of the demonstrators left the square and in acts of violence against property set alight cars, smashed windows and looted shops. It was a riot against the authorities and the symbols of affluence. *(continued)*

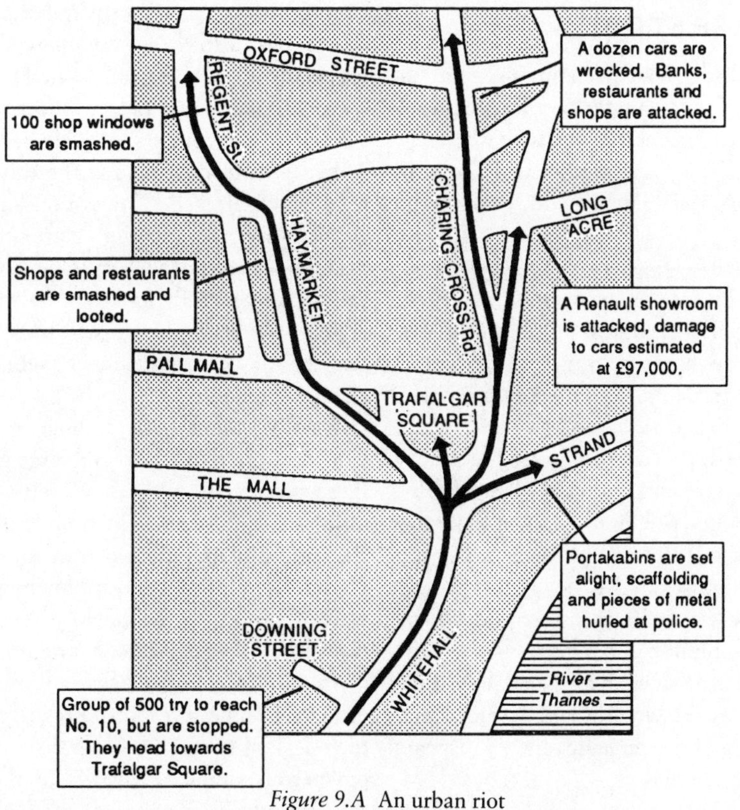

Figure 9.A An urban riot

There had been others. On 8 February 1886, there was a rally of the unemployed in Trafalgar Square. Over 600 police officers were present. A group of between 3000 and 5000 people went into Pall Mall, smashing windows in London's elite clubland. A year later another meeting of the unemployed in the square saw violent scenes as police and the army sought to regain social control. Two people were killed and over a hundred were injured.

Urban riots share a number of characteristics:

- they involve the occupation of, and struggle for control of, symbolic places. In London, Trafalgar Square is one such place.
- they include a confrontation between the forces of order and the power of demonstrators
- peaceful methods of crowd–police negotiation break down
- violence and law-breaking occur
- order is reimposed.

control over these things. Conflicts may occur when there are competing needs and interpretations. Some citizens want more localized, more accountable sources of power.

CASE STUDIES

So far we have discussed citizen involvement in general terms. In the remainder of this chapter I want to develop these general points with reference to specific case studies. These studies will allow us to see how urban social movements develop in specific contexts.

The south-east of England is the most affluent part of Britain. Unemployment rates are relatively low and economic growth rates comparatively high. Let us look at one particular region – Central Berkshire at one particular time, the early 1980s.

Location has been the key to this area's success (see Figure 9.3). Close to London, well served by good road and high-speed rail links it has the advantages of the metropolis but lower taxes for business, cheaper houses and easier access to the countryside. It is close to Heathrow airport which gives it good international connections, a vital requirement for firms moving high-value, low-bulk goods around the world. Growth and development pressure has taken three main forms: industrial, commercial and residential.

Central Berkshire lies within the M4 Corridor, an area which had the highest rate of high-tech employment growth in the country between 1975 and 1985, especially in the computer electronics sector. A survey by Berkshire County Council indicated that high-tech companies provided 14 per cent of the county's private-sector employment. There has also been commercial development, especially in Reading and Bracknell. The service sectors of insurance and producer services grew by 77 and 82 per cent respectively between 1971 and 1981, and Reading is now the third largest insurance centre after the City of London and Croydon. While office rents in Reading are similar to those in London (outside the City), rates (local property taxes) are generally less than half the level of the capital. Property developers thus receive similar returns on their investment, while users obtain cheaper premises.

Finally, there was a high rate of housing construction, particularly on large green-field sites. Effective demand was high because of the general affluence of the area, declining household size and the inflow of population from other parts of the United Kingdom. Local employment opportunities were plentiful in the healthy sectors, and good transport links facilitated commuting into London.

In summary, this was an area of absolute growth in some sectors, and relative growth within the national economy. Central Berkshire formed a prosperous suburban district of the metropolitan system. Jobs in traditional manufacturing indus-

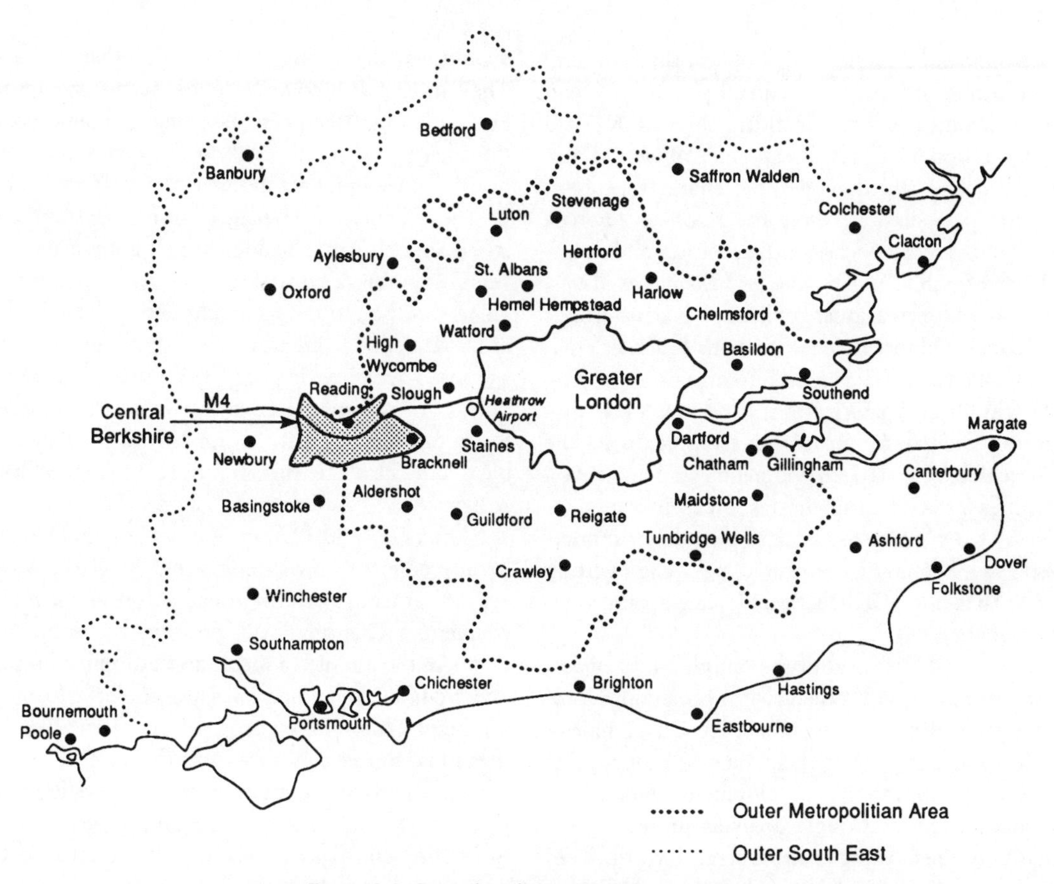

Figure 9.3 Central Berkshire in regional context

tries were lost here as elsewhere, but the growth in a number of service sectors and in high-tech manufacturing largely compensated for this decline, at least in aggregate terms.

Households as resident groups

Residents are affected by additions and changes to the built environment. The construction of a road or a new housing estate has both a direct impact upon the local environment and various externality effects, such as reducing local property values. In order to influence events, usually with the aim of maximizing positive externalities and minimizing negative ones, residents band together in resident groups.

How did residents react to this heavy development pressure? A study was conducted over the period 1981 to 1983 (Short *et al.*, 1986). A total of 149 resident groups in Central Berkshire were identified and contacted. Ninety-two of these were interviewed. Virtually all these groups were established after 1960, with increasing frequency during the 1960s, peak formation rates in the mid-1970s and fewer additions in the period to the early 1980s. This picture is consistent with increasing owner-occupation, the encouragement of public participation in planning issues and the rise of articulate middle-income groups; all factors related to resident groups' formation and action. The cumulative effect has been to generate a large number of groups seeking to influence planning outcomes.

All resident groups are concerned with their local environment. The term 'environment', however, needs careful inspection. Robson (1982) identifies three types of environment. The *physical environment* is, as the name implies, the form and distribution of buildings, roads, open spaces, and other features which constitute the natural and built environment. The *social environment* relates to the social and demographic characteristics of the local population, while the *resource environment* refers to the location, distribution and accessibility of public and private goods such as shops, schools and recreational facilities. Resident groups are concerned in varying ways with different environments. From the responses to questions probing the why of group formation, it became clear that groups wanted to prevent developments in the physical and social environments, and to obtain facilities or services in the resource environment. These two characteristics can be crudely summarized as stopping (or protecting) and getting (or enhancing). Three broad categories were identified.

Forming 43 per cent of the sample, the *stoppers* were primarily concerned with protecting local areas from further development. Just over half of these groups described their aims as stopping or modifying unwanted development while the remainder saw their main goals as protecting the quality of the existing environment. Two-thirds of *stoppers* were located in rural Central Berkshire, with the others in middle- and upper-income parts of the major towns. In 83 per cent of cases, group formation was initiated by a specific threat to the locality.

The concerns of *stoppers* reflect both environmental and economic calculations, and also specific social valuations. Residential growth is perceived to impose negative externalities in the form of increased noise, traffic congestion and construction activity, and leads to loss of land and landscape quality. But there is a hard-core of material interest underneath the environmental concern, relating to the impact of new development upon house prices. Moreover, many households have been attracted to certain villages, or the more salubrious urban neighbourhoods, because of their exclusivity. Particularly in the rural areas there is a powerful ideology which sees in a village location the hope of restoring a moral arcadia away from the anonymity of mass urban society. The concern with community here is an attempt to face up to modernity by asserting definable positions within a small local social hierarchy. New housing developments threaten this imagery.

The defence of villages from development comes mainly from residents of less than twenty years' standing who wish to maintain the physical village on the ground as much as the village in the mind. Defence is greatest when developments are proposed which either lead to the coalescence of villages or the submerging of a village by a larger town. Such changes strike not only at the material base but also at the emotional heart of the *stoppers*.

Getters, forming 37 per cent of the sample, were mainly concerned with enhancement of the local area in terms of both the social and resource environments. This involved pressing for improvements in the quality of local services, opportunities and facilities, and sometimes also took the form of self-help. Community participation, often allied to demands for community centres is seen as an essential part of community provision. The great majority (88 per cent) of these groups were found in recently constructed private sector estates, with only three (9 per cent) in areas of public housing.

It is the middle-income groups in 'average' areas who are the most vociferous in pursuit of improved services and facilities. A few may have close links with local political organizations and overlapping core membership; most, however, stress the non-partisan nature of their activities and campaigns. Self-help represents a step outside the formal political system when the state cannot or will not meet the demands of a locality.

There were some groups, 20 per cent of the sample, which were concerned in equal measure with the protection and enhancement of their locality. These *stopper-getters* were found both in old, established urban neighbourhoods and new estates. In only one case did a group's area constitute a predominantly public housing area, and in only three cases did it include both public and private housing. Owner-occupation dominated the other fourteen *stopper-getter* groups.

The questionnaire survey provides only a snapshot of resident groups in Central Berkshire and their roles. Group aims may alter over time, and thus a particular group will move between categories. For older neighbourhoods a typical progression is from *stopper* to *stopper-getter*, while for new estates the sequence may be from *getter* to *stopper-getter* to *stopper*. Group representing highly valued urban or rural areas are more likely to retain a basically anti-growth role.

In summary, the voice of the *stopper* is the voice of middle-class, middle-aged, owner-occupiers seeking to protect their physical and social environments. This voice is also strong in the other two types of resident groups, but it is not the only one. If you listen you can hear the sound of younger owner-occupiers in new estates and inner-city areas, and the demands of tenants' associations on council estates. Here the concerns are not only with protecting but also enhancing the local physical and social environments.

While it is relatively easy to note the aims, structure and external contacts of resident groups, it is very difficult to ascertain their impact. Between the goals of a group and final outcomes lie a myriad of conflicts and other interests. A precise balance sheet cannot be drawn up, but three general points can be made:

- the *stoppers* were influential in creating an articulate and powerful no-growth lobby in Central Berkshire, which has sensitized many local politicians to the issues of resisting and deflecting growth. The actions of the *stoppers* placed growth minimization higher than growth generation on the planning agenda of Central Berkshire.
- the importance of the *getters* increased. On the one hand they may become the *stoppers* of tomorrow. On the other, given their capacity for community self-help, they were least affected by reductions in certain public services. Given the drift of recent public policy toward control of expenditure and public service provision, the ability of communities to generate self-help creates further patterns of inequality. These distinctions will not simply be based on income and status, but also on length of residence – factors which create a strong sense of community within an area. With their success in mobilizing internal resources, and experience of campaigning for additional public provision, the *getters* may be best placed to counter some of the effects of public service decline.
- we can note that resident groups represent specific interests. Their primary concerns are the restriction of further growth and the generation of community facilities. They do not fully articulate the interests of the homeless or the unemployed and the voice of private sector tenants is scarcely heard. Not all issues which affect local communities are thus placed on the agenda for public discussion.

Conflict in the inner city

Located just east of London's financial centre, the 16 square miles of Docklands used to be the commercial water frontage of London (see Figure 9.4). It was also the home of working-class communities, almost 40000 initially based on dock-working. By the 1960s the docks were being closed because they were unable to cope with the bigger container ships. The port functions moved east to Tilbury and, in Docklands, registered dock employment fell from 25000 to 4100 between 1960 and 1981. Church (1988) provides a good review of the transformation of Docklands. This case study focuses on changes and tensions in this area in the late 1980s.

Close to the City, the area gave opportunities to developers for the modification of derelict land into offices and residences. There was an alignment of investment-rich institutions, a demand from a buoyant City for office property and new housing requirements of the growing new middle class, which all led to the modification and 'yuppification' of the area. The successful prosecution of these aims required three things:

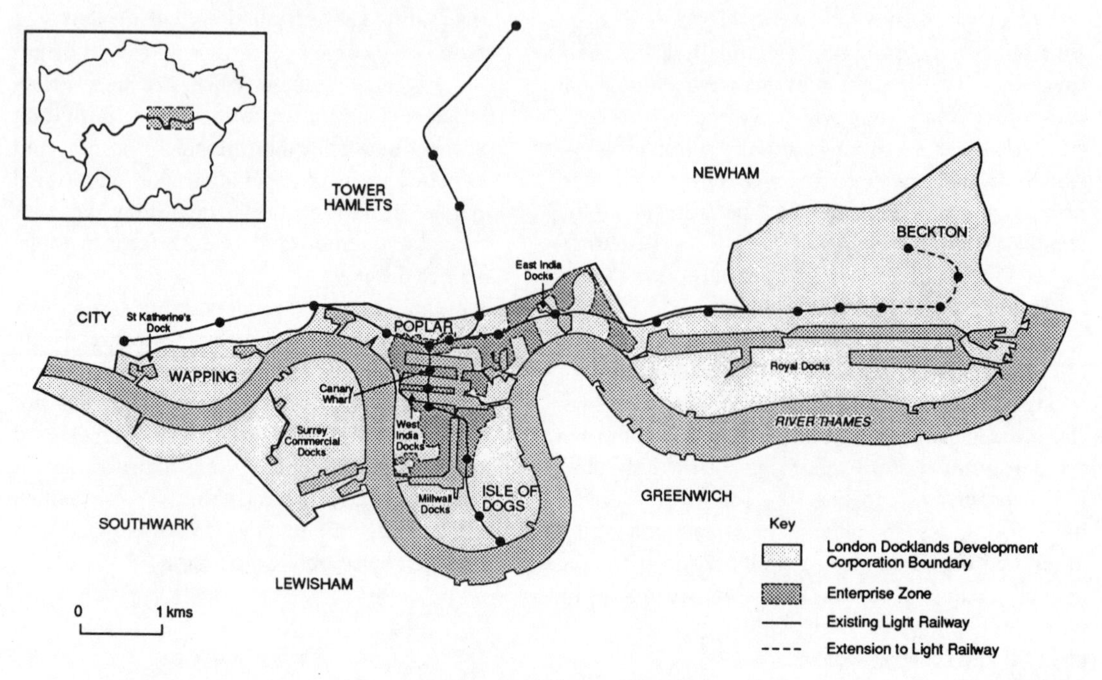

Figure 9.4 London Docklands

- incentives to private capital
- political power
- central organization

1 In the 1980 Budget the Chancellor of the Exchequer announced the creation of Enterprise Zones to promote private redevelopment of inner-city areas. Under this scheme incentives were provided over the period 1981–91 which included exemption from rates and land taxes on site disposal, tax allowances for building construction and relaxation of planning controls. Eleven zones were designated, one of them in Docklands, the Isle of Dogs (see Fig 9.4). This has been one of the most successful. For example, many newspaper offices have relocated from their traditional home in Fleet Street and now *The Sun*, *The Daily Telegraph* and *The Sunday Times* are located in the zone. Incidentally, their move was part of a restructuring of labour relations involving a reduction of the labour force and the introduction of new technology.

2 The commercial transformation of large areas will favour lucky landowners and astute developers but it will not directly benefit the local people. Any truly democratic local representation will thus tend to resist such changes. For the developments to take place, power must be taken out of local hands. This is the rationale behind the creation of the London Docklands Development Corporation (LDDC). The LDDC was established by a Conservative government in 1981. It replaced a committee, established in 1974, made up of representatives of five dockland boroughs. That committee was concerned with the needs of local residents. The non-elected, government-appointed LDDC has no need to court local political support. Its aim has been to 'develop' Docklands for the private sector.

3 Individual companies are unwilling and unable to undertake such large and speculative ventures. The LDDC has acted as central

> organizer of the project, assembling land, making environmental improvements and providing the vital, initial infrastructure investment. The LDDC spent £130 million between 1981 and 1985, almost £200 million from 1985 to 1987 including £35 million on a light railway system which links the area to the City, London's financial centre, and asked for £531 million for the period 1988 to 1993. The area is now an attractive location for office users, it is now 'closer' to the City and all of central London, yet rents are only a quarter of what they are in the City.

The LDDC was successful in raising awareness of the commercial opportunities of Docklands. A barrage of publicity changed the mental map of London. Previously the Docklands was unknown to the majority of middle-class Londoners. It was a spatially and socially self-contained segment of the capital. The LDDC campaign gave Docklands a higher profile and more 'positive' image. Publicity photographs were carefully taken to show only the glitzy areas, and colour enhancement changed the murky Thames into a sun-kissed, bright blue, pollution-free river. Almost £2200 million of private investment has been attracted. The whole area was transformed. Almost 500 000 square metres of office development are completed or under construction. At Canary Wharf was planned the biggest single development, a £3000 million complex of office and shopping space which was eventually expected to employ 72 000 people. The economic downturn and fall in commercial rents has hamstrung the whole development.

Housing has also been built. Thirteen thousand dwellings were completed or under construction by 1987. The LDDC plan is to complete 25 000 dwellings by the end of the century. Selling points have been the water frontages and the easy access to the City.

In effect there has been a transformation of the landscape of Docklands. The industrial buildings of the past are being recycled, both in terms of use and meaning. Docklands as Victorian economic resource is giving way to Docklands as postmodern landscape of offices, transformed from old working-class to new middle-class area. Docklands has become a showcase for the display of post-industrial employment and the presentation of housing forms for the newly affluent. The transformation of Docklands is not only a change in use but a change of meaning.

Two social forces are meeting in the same social space. On one hand there is *yuppification*, involving the destruction of an existing community and its replacement by a new middle class, with consequent changes in the meaning and use of space. On the other hand, there is *local resistance*. The press releases of the Isle of Dogs Neighbourhood Committee, for example, provide an antidote to the publicity machine of the LDDC. They point out that few of the jobs created have gone to local residents. When the average local income was £8500 per household the average price of a 2-bedroomed property in the area was £185 000. More radical has been the attitude of an organization called Class War. In its newspaper and billposters Class War urges local people to mug a yuppie, scratch BMW cars and make life as unpleasant as possible for the affluent incomers. Members of the TV soap *Eastenders* have been attacked, as Class War accused them of being show-business sellouts; estate agents regularly have their hoardings daubed with graffiti and 10 September 1988 was declared national Anti-Yuppie Day. Class War has 1000 members and a political philosophy. According to one spokesman:

> At first people thought we were just into violence. But we have our own political theory. We do not call ourselves anarchists any longer. Yes, we want to overthrow capitalism and if that has to be violent then so be it. We are interested in community politics, for the working class to stick up for itself.
>
> (Lashmar and Harris, 1988)

The threat is taken seriously by Scotland Yard who assigned six officers full time to monitor Class War in 1988.

Class War is unusual; more common are the

unorganized random acts of resistance/vandalism. As a correspondent in the *East London Advertiser* wrote:

> I was delighted the other day when I was sitting with my younger sister on the Isle of Dogs and saw some youngsters ripping up newly planted trees and using them to attack yuppie homes. Hopefully some young people locally will still have some fight in them and will repel these new Eastenders by making life unbearable for them.
>
> (Kane 1987)

As the letter suggests, young people constitute a point of resistance. They have energy, anger and have not yet learnt to accept their fate. In Docklands, however, it looks as if this resistance will ultimately fail. It is the deaththroes of a community undergoing marginalization and eventual disintegration. The organized Class War constitutes a nuisance and threat but not a permanent block to the changes. The power of finance capital in alliance with a central government committed to private enterprise and big business is too big an opponent for a small working-class community with few political friends and limited resources.

But the local youth still have power. Their very existence in the collective urban imagination has produced effects. There is the fear of crime. 'Colonization' of space involves the invasion of someone else's place. In the imperial past overseas colonization was underwritten by the British army and navy. Now it is the police who defend the urban colonizers. It is not that crime is any more prevalent in gentrified areas, although the contrast between rich and poor does provide greater opportunities. It is more a case of the new middle classes having the right language and the necessary confidence to demand better policing. Demands for more effective policing are greatest in areas undergoing gentrification.

The fear is also apparent in the new built forms. There is a contemporary urban enclosure movement which is blocking off and minimizing public open space. Riverside frontages are being alienated, walls are being constructed and barriers being created in order to keep out the urban folk devils. The security arrangements of residential blocks are a major selling point, while commercial properties are so designed that their frontages ward off rather than invite. The attraction of water frontages is only partly the scenic views, for on one side, at least, they can be easily defended against the urban 'other'. This bunker architecture is concerned more with security than display, personal safety more than show and the exclusion of indigenous communities rather than their incorporation.

The fear of the underclass has always been a major element in the life of London as in all world cities. In the past this has been managed by segregation, people knowing their place and staying in it. When different groups are in the same places the emphasis switches to the architectural design of the buildings, the location of those buildings and the construction of defensible spaces. In London's Docklands and selected areas of other world cities economic restructuring is causing a change of use, a change of meaning and a contest for the social control of urban spaces. The new urban order will arise from this struggle, its eventual shape a function of conflict and compromise, its final form a mark of victory. And of defeat.

Construction workers and the city

The configuration, use, size, internal layout and external design of the built environment embodies the nature and distribution of power in society; cities are systems of communications telling us who has power and how it is wielded. In this section I will look at the effect of one group on the structure of a particular city: the Builders' Labourers Federation (BLF) of New South Wales and the imprint of their actions on developments in Sydney in the 1970s.

The 1968–1974 property cycle

Sydney, like big cities in Europe and North America, experienced a rapid growth of office development in the late 1960s and early 1970s. The general reasons were a steady fall in the rates of return afforded to capital investment in manufacturing industry in the developed world at the same time as there was an increase in the amount of liquid capital. The declining returns afforded to manufacturing were caused by labour shortages and consequent militancy, raw material cost increases, and competition from the newly industrializing countries, with the precise mix varying in different countries. The growing pool of liquid capital was evident in the growth of financial institutions such as banks, pension funds, and insurance companies. At the international level, the collapse of the system of fixed exchange rates and the creation of an international finance market meant that capital could be shifted around the world. Dollars generated in Europe by US multinationals could be invested through London in commodity production in Asia or property development in Australia. Capital was invested in urban commercial property as there was a growing demand for office accommodation from the expanding service and financial sectors, especially in the large international cities. Commercial property was a favoured investment site because it took up large amounts of capital (an attraction for the big investment funds needing to place lots of investment), there was a demand, the investment was relatively trouble-free, and scarcity value was maintained by the absolute nature of space, reinforced in many places by planning controls.

The property boom also involved public authorities. The late 1960s and early 1970s was a time of increasing state expenditure. Expanded education and health programmes, for example, involved large building programmes. The state in its many forms (local states, statutory authorities, etc.) not only provided the context for property development and was an element in the demand but was also actively involved. Especially where authorities held land, the state could become a player as well as a referee in the property game. In some cases, the entry of public authorities politicized urban development issues even more, as questions of accountability and electoral liability were highlighted. State involvement brought into sharp focus the discussion about what type of cities were being produced, and for whom.

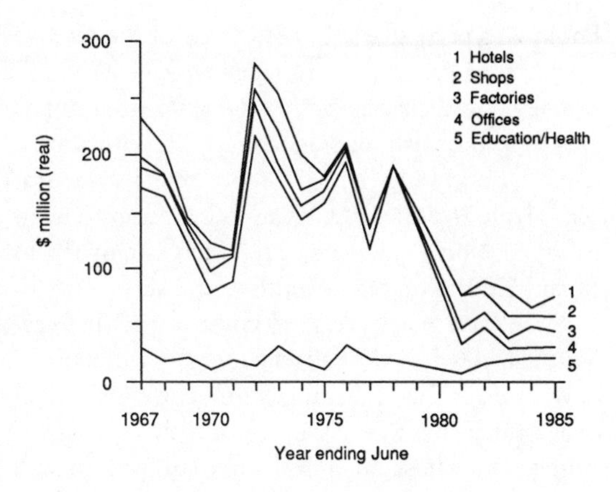

Figure 9.5 Value of non-dwelling building work completed in Sydney

The property cycle in Sydney

Some indication of the cycle in Sydney is given in Figure 9.5. Notice how the boom gathers pace in the late 1960s, peaks in 1971–2 and falls away to a relative slump in 1977. Thereafter there is evidence of another rise on the way. The boom in offices was fuelled by a number of factors. There was a growing need for offices from the expanding financial sector of banks, finance companies, and related businesses. The financial sector wanted offices in central locations, particularly in Sydney. The boom reflected and enhanced Sydney's pre-eminence over Melbourne as Australia's international city. There were also the requirements of a burgeoning public sector, which needed offices to house the expanding white-collar workforce and the specific buildings (for example, schools, universities, and hospitals) associated with its expanding functions. This period marked the coincidence of growth in both the private and public sectors.

The expansion of the commercial office sector was financed by both local and foreign capital. The boom was particularly large because of the relative openness of Australia to foreign investment. Traditionally, Australia, a dominion–capitalist country, was tied to British markets and linked with British capital, but during the 1970s investment came not only from Britain but from other countries in Europe, North America, and increasingly from Japan, Hong Kong, and South-East Asia.

Successive post-war state governments in New South Wales (NSW) had encouraged commercial development. The Liberal government which ruled over much of the cycle, from 1965–75, was simply the latest in a long line of pro-growth governments eager to attract capital investment. The Liberal government which came to power at state level in 1965 dismissed the Labour-controlled Sydney City Council in 1967 and appointed three commissioners. Between 1967 and 1969, the commissioners ran the city, allowing a flood of development approvals (see Figure 9.6). Thus, when the property boom reached its peak, there was a pro-development state government and a central business district (CBD) controlled by a non-elected, pro-growth triumvirate. After 1970 the City Council was an elected one, but dominated by the pro-business Civic Reform Association, which saw office development as a sign of metropolitan progress and, in some instances, a way of personal enrichment.

The boom: Before the boom, conditions in the construction industry reflected the power of capital and the weakness of labour. There was no holiday pay, few changing or washing facilities in what was a dirty job, no job security, and dangerous working conditions. The boom had two important features which affected capital–labour relations in the construction industry.

- Construction activity in central Sydney occurred mostly on large sites. The average size of development applications in the CBD of Sydney increased from 10 000 square metres in 1969 to 21 000 square metres in 1974. For labour, the large sites meant more workers, longer work life on the sites, and thus easier conditions for organizing the workforce.
- Many of the building projects were speculative enterprises. In the public sector projects there was a specific client eventually responsible for picking up the bill, but the private sector projects were built to meet a perceived demand, not the requirements of specific clients. Development companies needed to borrow money to finance their operations. Credit allowed the projects to be built but made the companies vulnerable. Credit lines could not be extended indefinitely.

In general, the building boom gave extra leverage to labour: their strength was enhanced by the low levels of unemployment which between 1965 and 1975 did not go above 3 per cent for Australia. In effect, the property boom meant a potential shift in the power relations between building capital and building labour.

The building boom meant more workers on more easily organized big sites. The BLF increased its membership from 4000 in 1968 to 10 000 by 1970. It thus became easier to organize strikes (see Figure 9.7). In 1970, after the failure of negotiations with the Master Builders' Association (MBA), the BLF in New South Wales organized a five-week strike in support of increased wages and better conditions. In the third week of the strike, the BLF adopted a selective strategy. Companies which signed the agreement for improved wages and conditions would have the bans lifted. On 29

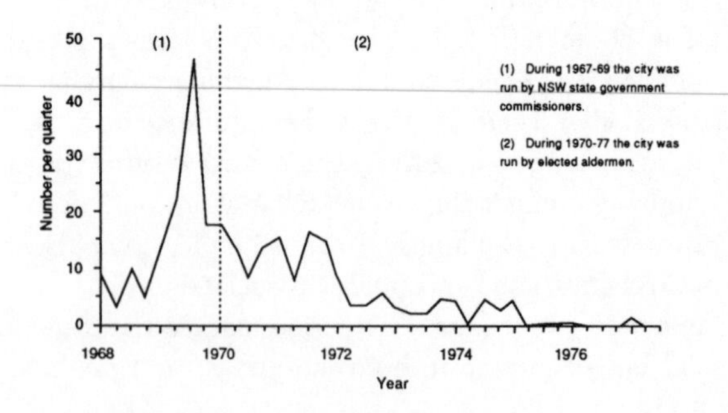

Figure 9.6 New building development approvals in Sydney CBD

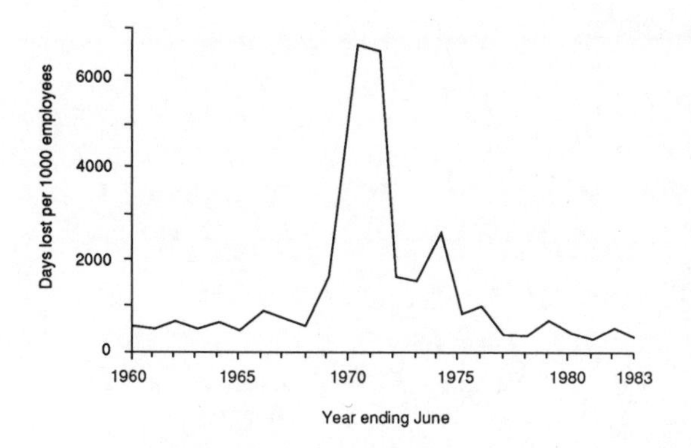

Figure 9.7 Working days lost per 1000 employees in the New South Wales construction industry

May 1970, five major firms signed the agreement. The BLF successfully exploited a major cleavage in the construction industry. Although all companies may benefit from group solidarity, individually it is more rational for them to meet workers' requests in order to finish the job. As the Director of Industrial Relations of the MBA noted in an interview with me in 1986: 'If you're not working, it costs you between $30 000 and $50 000 in interest payments. So when unionists ask for more money amounting to, say, £3000, well there is no comparison. £3000 versus $50 000'. Two weeks later, the MBA agreed to go to arbitration. The BLF achieved its objectives. The strike had been successful, gains had been made, and building workers' confidence was high. The strikes successfully gouged out benefits from the profits of the building boom.

Green Bans: Organized building labour and the BLF, in particular, also extended their concerns beyond the narrow range of wages and conditions of employment in what came to be known as 'Green Bans'. These involved union actions to block development, not for increased wages, but because of environmental and wider social considerations.

In the early 1960s, the BLF had been a right-wing union whose leadership seemed to have little concern with improving the lot of the workers. A radical rank and file group within the Union sought to gain control and eventually were successful in the 1964 union election. For the next ten years, the unions followed a radical line dominated by notions of direct action, rank and file participation, regular elections for union office, and the fostering of community–labour links. The leaders, Joe Owens (Secretary, 1973–4), Jack Mundey (Secretary, 1968–73), and Bob Pringle (President, 1969–75), were committed to improving conditions in the industry as well as to wider political goals. Mundey was arrested in anti-Vietnam war demonstrations, and in 1971 Bob Pringle and another BLF member cut down the goalposts at a ground where the Springbok rugby team were to play. The leadership shared the ideology and had adopted the tactics of direct action of the New Left, then developing in Australia as well as North America and Europe, particularly around the issue of the Vietnam war. Green Bans were part of a wider political struggle and a broader political philosophy.

The first Green Ban began in the upper/middle-income Sydney suburb of Hunter's Hill. In 1970, a plan to build fifty-seven townhouses on a piece of open space known as Kelly's Bush met with local resistance. The Battlers of Kelly's Bush were a group of local residents, all women, who sought to resist the development and maintain the open space. The Battlers wrote to local and state governments but met with little success. In 1971 they were approached by the BLF leadership. The BLF pledged their support and put a ban on any construction work at Kelly's Bush. For the next few years, during the peak of the building boom, the BLF Green Ban policy was a major factor in shaping Sydney.

There were two types of ban. Permanent bans involved a resolute commitment to a particular action. Examples included the refusal to work on proposed plans to build an Olympic Stadium in Centennial Park, the refusal to demolish the Theatre Royal, and the Green Ban put on the proposed underground car park opposite the Sydney Opera House because it involved the destruction of old fig trees. Temporary bans were used to give greater strength to groups (usually residents) in their negotiations with developers and state authorities. Temporary bans were also used to aid other groups. In 1973, for example, the BLF imposed a ban on work at Macquarie University because a homosexual student had been expelled from one of the residential colleges. In the same year, a ban was imposed on the University of Sydney because two women were not allowed to give a course on women's studies. In both cases, the BLF action led to further negotiations and eventually the student was reinstated and the course was given.

The Green Bans involved local resident groups. There were good reasons for resident group activity. The building boom had affected the residents of inner Sydney; they faced incursions into their open space, the negative externalities of new building, increased traffic, and, for the lower-income groups, reduction of housing opportunities as developers sought to build offices and expensive apartment blocks. The local planning system gave almost no voice to local residents. In 1971, the Coalition of Resident Action Groups (CRAG) was established in Sydney with the aim of exchanging information, organizing joint action, and general lobbying. By 1972, CRAG and the BLF were in an alliance. There were advantages for both sides. For CRAG, an alliance with the BLF gave them the power to stop developments, because the BLF had the muscle to stop demolition and halt building projects. For the BLF, CRAG and its members gave legitimacy. Throughout the Green Ban period, the BLF faced heavy criticism from developers, the state authorities and the press that their actions were undemocratic. The Sydney press was savage in its attack. The *Sydney Morning Herald* (14 August 1972) wrote about 'delusions of grandeur' and 'the highly comical spectacle of builders' labourers ... setting themselves up as arbiters of taste and protectors of national heritage', and *The Australian* in an editorial of 3 September 1972 noted:

> When the vocal leader (Mundey) of a tiny minority in one union begins to sway public and municipal decisions on multi-million dollar questions in which he has no expertise whatever, it is time to begin asking what has gone wrong with the process of government in this country?

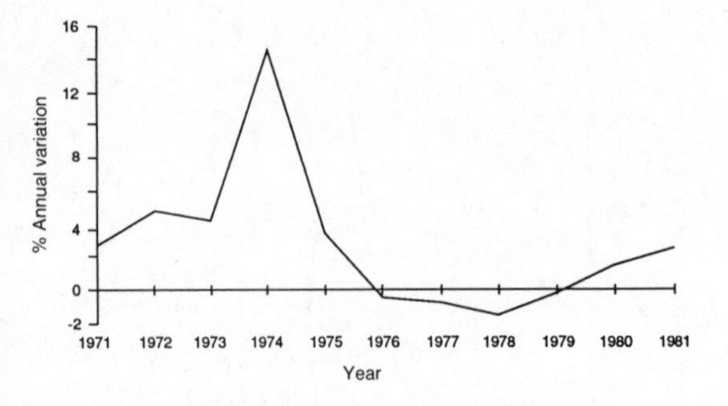

Figure 9.8 Variation in wages and earnings of building workers, net of the consumer price

The press campaign personalized the issue, focusing on Mundey and his membership of the Communist Party of Australia. The assumption was that mere labourers had no role to play in city planning; they were communists who were endangering big projects and frightening away foreign investors. The resident group connection showed that the BLF had wider support. The policy of imposing bans only when there was local resistance ensured and reflected popular support.

The slump: By 1973 the building boom was beginning to slow. In November 1973 the Australian Financial Review had a headline of 'Property Sales Bubble Bursts'. In the same year, there were over half a million square metres of office space in Sydney CBD lying vacant, a fifth of the total office space in the city. By 1976 the boom was over and unemployment was beginning to rise. Building workers were affected in terms of growing unemployment and falling wages (see Figure 9.8). The slump meant a decline in the demand for labour and a weakening in the bargaining power of the building unions. In effect, the slump shifted the balance of power away from organized labour towards capital.

The growth and decline of the Green Ban movement can be seen against the background of the building boom. For the workers in the BLF, the boom meant secure continued employment. As one right-wing BLF member remarked to a journalist in 1972:

> Our members tolerate Mundey's views because they really don't matter. Times are easy for us: there's plenty of work around. We get pretty good money and if Jack Mundey wants to sprout [sic] off about things, that's OK. But things would be different if things got hard in the industry.
> (*The Australian*, 5 September 1972)

It is too easy, however, to see the Green Bans simply as a function of the building boom. There had been booms in the past without Green Bans and not all building unions at the time pursued a Green Ban policy. The BLF leadership made a successful connection between working conditions, rates of pay, and broader social and environmental issues. The radical leadership won the confidence of the rank and file through its pursuit of higher wages, its policy of limited tenure of office, and its openness to bottom-up policy making. The radical BLF showed that alliances between labour and residence-based urban social movements are possible and that connections between production-based and consumption groups are feasible. The circumstances have to be right but a crucial element is the existence of people with the vision to make the connections.

The demise of the radical BLF and the decline of the property boom meant a setback for oppositional movements in Sydney. But the story is not one of increasing gloom. We can identify at least three enduring positive consequences of the 1968–74 boom period.

- not all the gains of construction labour were lost in the subsequent slump. The Green Bans were dropped, but the improvements in working conditions did not revert to the pre-boom position. The almost total unionization of big sites continued and a closed shop of 'no union ticket–no start' was effectively established for all large non-residential sites. The issue of safety, once raised, refused to disappear. Most sites now have safety codes, and safety considerations have been incorporated into design criteria. All high-rise buildings in Sydney now have safety nets. The boom allowed organized labour to civilize much of the industry and put capital–labour relations on a new terrain of conflict and compromise. Agreements in 1988, for example, included a 38-hour week, a portable superannuation and long-service leave schemes, and a national safety code. There is a tension in this relationship. For the union representatives, especially of the larger unions, there is a danger of incorporation, of putting claims of particular sites into line with broader coporatist deals. As for the builders, they want to deal with only a few union officials because this ensures easier negotiations, but it concentrates power in the hands of those few officials. Organized labour wants power without too much responsibility whereas capital wants the unions to have responsibility without too much power.
- the struggles of the BLF and community groups such as CRAG sensitized a broader public to environmental issues. The Labour government of New South Wales which came to power in 1976 established a Land and Environment Court, a Heritage Council, and in various environmental planning Acts sought to incorporate public participation. The spirit of legislation, if not its practise, owes much to the Green Bans.
- the social struggles of the period have an enduring legacy on the landscape of Sydney. The fig trees still grow opposite the Opera House, Centennial Park does not have an Olympic Stadium (yet) and many of the new developments are retaining the facade of buildings previously scheduled for demolition during the boom. These are not once-and-for-all victories. Consider The Rocks, one of the oldest parts of Sydney adjacent to the CBD, saved in the 1970s from high-rise commercial development in a bitter struggle between the BLF and residents on the one hand and devel-

opers and the state government on the other. As demand for more office space continues so the pressure builds up on The Rocks. Old terraces are being turned into offices, up-market and tastefully renovated, but offices all the same. The old buildings are being retained but there is a change in the social community if not in the physical facades. Parts of The Rocks are becoming tourist centres, the site of a commodified history of charming old Australia and a classless Sydney. But it is not yet an outright loss. In 1984, Mundey was elected alderman to Sydney City Council by people in The Rocks. From this base, he fought against commercial expansion in the area. And at weekends, and Friday nights in particular, The Rocks continues to be an entertainment centre for ordinary Sydney-siders. The streets, full of Japanese and American tourists during the day, at night resound to the extended vowel sound of young Sydney-siders still given a place to play in their city.

GUIDE TO FURTHER READING

For general approaches to urban social movements have a look at:

Castells, M. (1978) *City, Class and Power.* Macmillan, London.

Castells, M. (1983) *The City and The Grassroots.* Edward Arnold, London.

Lowe, S. (1986) *Urban Social Movements.* Macmillan, London.

Pinch, S. (1985) *Cities and Services: The Geography of Collective Consumption.* Routledge & Kegan Paul, London.

Saunders, P. (1979) *Urban Politics: a Sociological Interpretation.* Hutchinson, London.

The three case studies in this chapter are drawn from my more detailed work:

Short, J. R., Fleming, S. and Wipp, S. (1987) 'Conflict and compromise in the built environment', *Transactions, Institute of British Geographers N.S.*, 12, 29–42.

Short, J. R. (1988) 'Construction workers and the city', *Environment and Planning A*, 20, 719–40.

Short, J. R. (1989) 'Yuppies, yuffies and the new urban order', *Transactions, Institute of British Geographers N.S.* 14, 173–88.

Short, J. R., Witt, S. and Fleming, S. (1986) *House-building, Planning and Community Action.* Routledge & Kegan Paul, London

Relevant journals

Environment and Planning, A, C and D
International Journal of Urban and Regional Research

Other works cited in this chapter

Church, R. A. (1988) 'Urban regeneration in London Docklands: a five year policy review', *Environment and Planning C*, 6, 187–208.

Dahl, R. A. (1961) *Who Governs?* Yale University Press, New Haven

Hayes, E. (1972) *Power Structure and Urban Policy: Who Rules in Oakland*? McGraw-Hill, New York.

Hunter, F. (1953) *Community Power Structure.* University of North Carolina Press, Chapel Hill.

Kane, F. (1987) 'The new eastenders', *The Independent*, 26 September.

Lashmar, P. and Harris, A. (1988) 'Anarchists step up class war in cities', the *Observer*, 10 April.

Lukes, S. (1974) *Power: A Radical View.* Macmillan, London.

Robson, B. T. (1982) 'The Bodley Barricade: social space and social conflict,' in K. R. Cox, and R. J. Johnston, (eds) *Conflict Politics and The Urban Scene.* Longman, London.

Strong, J. (1885) *Our Country: Its Possible Future and Its Present Origins.* Baker & Taylor, New York.

POSTSCRIPT

This book has covered such a rich variety of issues, concerns and questions that there seems little point in providing a summary conclusion. What is the point, after all, of writing nine chapters, to then compress the detail into a dense, final chapter? However, the very variety suggests the need for some kind of concluding statement. There are a number of consistent strands and recurring patterns that run through this book. I will consider the most important ones.

This postscript also gives me an opportunity to look at possible future outcomes of present trends. I realize the risks. Between the first edition of this book and the second, tremendous changes have taken place. Changes which could not have been predicted. It is difficult, therefore, to make any concluding statements with any real measure of confidence. The 1980s has been a salutary experience for those who would predict the future. The 1990s may prove equally destructive of long-term projections based on existing states or current affairs.

THE NEW WORLD ORDER

As the 1980s came to a close the term *new world order* was heard more often. In essence it referred to the decline of the old dichotomy between East and West, the decline of the Soviet Empire and the unravelling of state socialism in Eastern Europe. For many, this was heralded as the dawn of a new era. In many ways it is. The old bipolar division which structured the world order is coming to an end. By the time you read this book it probably already has. The USA–Soviet enmity will, like that between Rome and Carthage, dissolve into history. That is something to be applauded. The build-up of arms, the repression of dissent, especially behind the Iron Curtain, and the support of authoritarian regimes by both superpowers around the world did much to slow down, if not halt, social progress. But the New World emerging from the Old World is not free from tension; we can identify two major sources.

First, there is still a huge disparity between the rich countries of the world and the poor. The gap between the haves and the have-nots is still there and in some cases is increasing. Even in the middle-income countries which had such tremendous growth rates in the 1970s the debt crisis is gnawing away at the very fabric of society. In countries such as Mexico and Peru rates of malnutrition actually increased in the 1980s. The global economy is still unfair in its distribution of costs and benefits. The net effect is that a small proportion of the world's population suffers from '*affluenza*' – problems associated with too much wealth, too many choices and too little meaning in their lives – while a significant proportion has difficulties in sustaining the basic essentials of a decent life for themselves and their families. As we reach the end of the second millennium there are still too many hungry people in the world.

Second, in the rich core, there is increasing economic competition between former political allies. The conflict between Japan, the USA and Europe was masked in the early to mid 1980s because of economic growth, but as the world economy hit recession in the early 1990s it became obvious that erstwhile allies were fighting an intense war of economic competition. The decline of the Soviet Union means that there is no longer a political 'other' to provide the cement to bind these countries together. Conflict between them will take the form of attempts at protecting their domestic markets, fighting for overseas markets, and emerging conflict concerning the cost of international policing. The USA has borne the burden of military expenditure amongst Western powers, but with its own economic difficulties and the decline of the Soviet Empire, politicians and the electorate in the USA may well ask why the defence burden should be so unfairly distributed, especially given the economic strength of countries such as Japan and Germany.

We can summarize the picture thus: from 1945 until the mid 1970s the Western World was dominated by the USA, which was the strongest economy and the biggest military power. There was a large measure of congruence in the foreign policy of the allies. Throughout the 1970s and 1980s the USA remained the leading military power but its economic strength was weakened by an emerging Europe and a dynamic Japan. The 1990s will see the creation of competing trading blocs: an integrated Europe with Germany as the leading economic power, Japan with a worldwide presence but with a very strong presence in Asia and the whole Pacific Rim, and an American bloc consisting of both Latin and North America with the USA as the dominant power. These three blocs will seek to protect their domestic industries and ensure export markets. The stage is set for an era of intense international economic competition.

Global concerns in the new world order

The term, 'new world order' has an air of optimism, a sense of new opportunity. Perhaps that is why it has such vogue for a while; it gave the possibility of hope. In chapter 8 we considered the concept of the global village. For the global villagers of the 1990s there are a number of causes of concern as well as sources of optimism. Let me mention two of the most important.

The fundamental importance of environmental issues

Environmental issues have always been discussed. More recently, there was the strengthening of the environmental movement in the late 1960s and early 1970s. The world recession of 1974, however, shifted attention away from the environment and onto issues of economic growth. The 1980s witnessed a revival of significant interest in environmental issues, which, I think, will be of lasting importance. The concern with the environment is no longer the preserve of middle-class intellectuals in the affluent suburbs of the rich world, it goes wider and deeper. It links people across the world. There is now an awareness that what happens in one part of the globe, be it burning of rainforests, an explosion at a nuclear power plant or the emission of car exhaust fumes, has a direct impact on all the other parts. There is a web of ecological processes that makes us dependent on the natural resources of the world and on the actions of fellow global villagers. There is now a whole series of issues – protecting the rainforest, maintaining ecological diversity and ensuring the continued 'livability' of the planet – which transcends national boundaries and the concerns of just the rich world. There is an acceptance of the tremendous importance of the ecological link between the human and the natural world; this realization will inform world debates and national politics for many years to come.

War and peace

Questions of war and peace have not disappeared with the decline of superpower rivalry. In some ways, to be sure, the world is a safer place now than it was when the USA and USSR had huge nuclear arsenals and their policies seemed to be predicated on the bizarre and crazy assumption that nuclear holocaust was an acceptable piece of military strategy. Thankfully, that scenario has disappeared. However, nuclear proliferation is increasing and now almost a dozen countries have the capacity to produce nuclear weapons, and not all of them are bastions of freedom and democracy. The bipolar world had a kind of mad stability while the new world order has a dangerous degree of instability and anarchy. The peace movement of the 1980s was influential in shifting world opinion away from a posture of attack toward a more conciliatory approach. The peace activists of the 1990s have a more difficult job: to persuade public opinion that the world is still a dangerous place, in some cases more dangeous, and that as long as states continue to pursue policy objectives through military means, war, destruction and death will threaten the globe.

The state

The state is the point of connection between the world order and the lives of ordinary people. The state links the global economy with the household economy, space and place, the global and the local, the generality of world order with the particularity of individual households living in specific places. In chapter 4 we considered the concept of legitimation crisis. This was defined as the inability of the state to secure popular approval. Legitimation crises arise for a number of reasons. We can identify two major ones in the contemporary world.

Boundaries of state and nation

In many parts of the world there is still a mismatch between the boundaries of the state and those of the nation. The Soviet Empire has fractured along national lines and the post-war states of Eastern Europe are subject to centrifugal forces as old nationalities emerge. As a result, a few countries, the prime example being Germany, have seen increasing congruence between nation and state, while in most others the birth of new nations has proved a highly explosive issue as old nations emerge from newer and different state boundaries. Nationalism continues to exercise the popular imagination and provides an important vehicle for the mobilization of popular protest. The continual drive to national expression provides a source of major political change in the world. As long as state boundaries fail to express national identities there will be a source of dispute between parts of the population and the operation of the state.

Throughout most of the twentieth century it was assumed by many that nationalism was an old-fashioned concept with little place in the modern world. Indeed, there were whole ideologies and political movements, socialism being the most important and persuasive, that were ostensibly based on the end of nationalism and the demise of the parochial and limited interests of the contemporary state. However, nationalism has proved to be more resilient. More than that, it seems to be growing. In Eastern Europe, for example, it is re-emerging from the confines of state socialism which preached the message of universal brotherhood, even if it did not practice it.

The renewal of the nationalist enterprise is part of the broader move, the shift from modernism to postmodernism. Modernism was concerned with space, with universals, with the forward march of history. Postmodernism is concerned with place, with local knowledge and particular identities. Modernism looked forward to a more uniform world, postmodernism revels in the variety of the world. A postmodern world is one where identity is based on a hierarchy of levels, not just the global as with the modernist conception, but global *and* national *and* local *and* community.

State irrelevance

The state in the modern world is caught in the pincers of irrelevancy. There seem to be two trends. The first is the move toward larger groupings of states. There is a variety of alliances, including the economic groupings which we have already referred to, such as the EC, and there is the United Nations which provides a forum for all the states in the world. This trend is associated with the increasing perception that the major problems and issues which face us are global problems and world issues. Environmental degradation, the fear of war, the obscenity of starving children, these are all things that can only be resolved through international action and co-operation. The state, on its own, cannot adequately address, never mind solve, these problems.

The second trend is a concern with the local, an awareness of the importance of place, an identity with community. National space is often too large to reflect national identity or the significance and meaning of community.

The outcome is that the state, at its most awkward, is too small to address global issues yet too big to respond to local concerns. The state is too parochial to meet the needs of a global community yet too big and distant to meet the requirements of a local community. The state will continue to exist – where else can politicians go and what else can they do – but as a source of emancipatory change the state is becoming more and more irrelevant.

Social movements

In Part III we looked at the active role of the population in social and political changes. This bottom-up view was a necessary corrective to the predominant top-down perspective of political geography. A variety of movements were considered. Here, we will mention briefly the major social movements of the 1990s and beyond:

The politics of environmental concern

The politics of environmental concern will become a dominant movement at global, national and local levels. Green politics will be allied to a range of other concerns, including the women's movement, the pacifist movement and the rights of indigenous peoples. A whole set of concerns will cluster around and find coherence in the notion of environmentalism. The questioning of economic growth as a national priority, quality of life issues and the concern with enabling everyone to lead a dignified, sustainable life will all be addressed in and through the environmental question.

Global fairness

The concern with global fairness will be an important agenda for the 1990s. The disparities in the quality of life of different peoples will continue to haunt the imagination of the wealthy and the concerned. We can picture a continued attempt, be it in fund raising, education or some form of public service, to redress the global imbalance in life chances. Not everyone will care and even those who do will not all take action, but there will be a significant number who are both concerned and active. Global justice will provide one of the few beacons to guide the concerned, the guilt-ridden and the activist of the 1990s.

Citizens and the city

At the local level, citizens will concern themselves with the universal issues of getting good jobs, decent housing and the proper range of public services to ensure a good life. The definition of a 'good life' has changed and continues to vary around the world. We can see the change most obviously in the cities of the rich core where citizens no longer want just any job. People are as concerned with the quality of employment as the quantity of employment. In the 1990s, as before, those who are struggling will perhaps take any available job, but more and more people are concerned with finding employment that not only pays well but is socially useful, ecologically respectful and allows them to enjoy the non-working hours. As economic growth and income maximization have to compete as social and personal goals with ecological responsibility and maximization of the quality of life, the nature of citizen activity will change. People will become even more concerned with the quality of public services and after the private greed of the 1980s public responsibility will become an important social objective. The fundamental question is, can the rhetoric be turned into reality? That will be the battle for the citizen activists of the future.

In the poorer cities of the world most citizens do not yet have the luxury of choice. For many, the basic necessities are still the main goal in life. But we would do well to look at the struggles and successes of Third World citizens with more than just patronizing interest. For too long the aim has been to export techniques and technology from the rich world to the poor world. The results have been ludicrous: motorways built for cities where most people are too poor to afford private cars; or downright dangerous, as in encouraging mothers to feed their children inferior powdered milk rather than the more nutritious breast milk. A more radical solution, and one which ties in with earlier comments about a simpler, more ecologically sound future, is to look at the self-help strategies and low-tech solutions used by the citizens of many Third World cities and to see them as prototypes for more universal application. This is not to preach a low-grade urban environment but to suggest that we can learn something from the success stories of people in poorer cities who have realized that ultimately the most important resource of any city is its citizens.

The world is in flux. Old empires are falling, new states are emerging. There is a sense of profound change. The past no longer provides us with a secure guide to the future and even the present is difficult to comprehend. At times like these we should consider our knowledge provisional, limited and based on events which may no longer have relevance. The lessons of the 1980s, for me, were twofold. First, a sense of humility as even the securest foundations of the world order were swept aside and, secondly, that there is an important role for ordinary people in changing structures and influencing events. These are profound lessons for the 1990s ... and beyond.

INDEX

WITHDRAWN